Differential Forms

with Applications to the Physical Sciences

Harley Flanders
University of Michigan, Ann Arbor

DOVER PUBLICATIONS, INC.
New York

Published in Canada by General Publishing Company, Ltd., 30 Lesmill Road, Don Mills, Toronto, Ontario.
Published in the United Kingdom by Constable and Company, Ltd., 10 Orange Street, London WC2H 7EG.

This Dover edition, first published in 1989, is an unabridged, corrected republication of the work originally published in 1963 by Academic Press, Inc., New York. Please see the Preface to the Dover Edition on page ix for an explanation of revisions made in this text.

Manufactured in the United States of America
Dover Publications, Inc., 31 East 2nd Street, Mineola, N.Y. 11501

Library of Congress Cataloging-in-Publication Data

Flanders, Harley.
Differential forms with applications to the physical sciences / Harley Flanders.
p. cm.
"Unabridged, corrected republication of the work originally published in 1963 by Academic Press, Inc., New York"—T.p. verso.
Includes bibliographical references.
ISBN 0-486-66169-5
1. Differential forms. 2. Mathematical physics.
I. Title.
QA381.F56 1989 89-36936
515'.37—dc20 CIP

To June

Foreword

After several friendly discussions of the pros and cons of tensors versus differential forms in the solution of engineering problems, I persuaded my colleague Dr. Flanders to prepare a number of lectures on differential forms. The result was an outstanding series of lectures which was presented to a group of interested faculty members within the several schools of Engineering at Purdue University.

It became obvious to those attending that the use of differential forms would give them another tool for the analysis and synthesis of engineering systems. There are certain problems, normally very difficult to solve by using tensors only, for which results are more quickly and directly obtained with differential forms.

The author was encouraged to formalize his notes to the extent necessary for publication, to enable others to study this important subject. The text is recommended highly because differential forms and related concepts which have evolved from modern mathematics are new and powerful analytical tools for use by the engineer and scientist.

George A. Hawkins, Dean
Schools of Engineering and Mathematical Sciences
Purdue University
November 20, 1962

Preface to the Dover Edition

I have made the following changes to the 1963 edition. First, I have rewritten the proof of the Third Lie Theorem, starting on page 109, to better systematize the computations. Second, I have rewritten the derivation of some basic relations in phase space, pp. 164–165, to be less computational and more in the spirit of the book. Finally, there is an addendum to the bibliography on page 199 which should be useful. In particular, it mentions the forthcoming MAA Studies volume in which I give Kannai's differential form proof of the Brouwer fixed point theorem.

HARLEY FLANDERS

June 1989

Preface to the First Edition

Last spring the author gave a series of lectures on exterior differential forms to a group of faculty members and graduate students from the Purdue Engineering Schools. The material that was covered in these lectures is presented here in an expanded version. The book is aimed primarily at engineers and physical scientists in the hope of making available to them new tools of very great power in modern mathematics. Although none of our applications goes very deep, it is hoped, nevertheless, that enough ground is covered in each case to indicate the usefulness of this machinery.

A word about the organization of the book is in order. The first chapter is introductory and sketches where we are going and why. Chapters II, III, and V include all of the theoretical material; a knowledge of this opens the door to the applications. Probably on first reading, one should aim more at developing some intuition for the subject and getting a firm idea of what the various different things which are defined look like, rather than at working out proofs in detail. Applications to questions in differential geometry (including many topics of considerable use in physical sciences) are mostly in Chapters IV, VI, VIII, and IX. Applications to various topics in ordinary and partial differential equations will be found in Chapter VII. Finally, applications to several topics in physics are in Sections 3.5, 4.6, 6.4, and Chapter X.

What is presupposed of the reader is first of all a certain amount of scientific maturity, the precise direction not being too important. While the book is not really advanced mathematics, it is not exactly ground floor mathematics either, and a reasonable knowledge of the calculus of functions of several real variables is necessary, as is a working knowledge of linear algebra through the ideas of linear combination, basis, dimension, linear transformation. Some exposure to a minimum amount of the ground rules of modern mathematics, sets, cartesian products, functions on sets, is helpful but not essential. This material is usually picked up by osmosis anyway, and the Glossary of Notation at the end of the book should be helpful. The reader should also know about the existence of solutions of ordinary differential equations. A passing familiarity with tensor methods is useful, but not essential.

If our audience consisted of mathematicians alone, it would be in order to use somewhat more care in our formulations of definitions and proofs of theorems and to discuss in considerably more depth numerous technical points we here pass over lightly. Our goal, however, is to develop an intuition and a working knowledge of the subject with as much dispatch as is possible. This perhaps could be done in less space except for our insistence on a degree of rigor matching that found in the better treatises on theoretical physics. This falls short of the extremely great precision which is customary in modern abstract mathematics and pretty much inherent in its nature. One who quite rightly is searching recent developments in mathematics for applicable material must find this precision a considerable barricade, overpedantic if not downright tedious—a very real factor in the great separation between modern mathematics and modern science. Making his craft available to science is not a light task for the mathematician and the extent to which this book makes a contribution therein must necessarily be its primary measure of success.

In spite of all this, we do not hesitate to recommend this material to graduate students in mathematics as an introduction to modern differential geometry; indeed, a well-trained advanced undergraduate should find the book quite accessible. Considering the degree to which present day mathematical training consists of one abstraction after another, some of the things in this book could be a bit of an eye-opener, even to a mathematics student who is well along. For example, one could envisage such a student meeting here a parabolic differential equation, or a matrix group, or a contact transformation for the very first time.

It is my pleasant duty to acknowledge the substantial help and encouragement I have always had from my teachers, colleagues, and students. In this respect a special vote of thanks is due George A. Hawkins, Dean of the Schools of Engineering and Mathematical Sciences of Purdue University. Finally, I wish to express my gratitude to Elizabeth Young, whose beautiful typing of the manuscript was a substantial contribution.

July 1963 HARLEY FLANDERS

Contents

V. Manifolds and Integration

VI. Applications in Euclidean space

VII. Applications to Differential Equations

VIII. Applications to Differential Geometry

IX. Applications to Group Theory

X. Applications to Physics

I

Introduction

1.1. Exterior Differential Forms

The objects which we shall study are called *exterior differential forms.* These are the things which occur under integral signs. For example, a line integral

$$\int A\,dx + B\,dy + C\,dz$$

leads us to the one-form

$$\omega = A\,dx + B\,dy + C\,dz;$$

a surface integral

$$\iint P\,dy\,dz + Q\,dz\,dx + R\,dx\,dy$$

leads us to the two-form

$$\alpha = P\,dy\,dz + Q\,dz\,dx + R\,dx\,dy;$$

and a volume integral

$$\iiint H\,dx\,dy\,dz$$

leads us to the three-form

$$\lambda = H\,dx\,dy\,dz.$$

These are all examples of differential forms which live in the space $\mathbf{E}^3$ of three variables. If we work in an n-dimensional space, the quantity under the integral sign in an r-fold integral (integral over an r-dimensional variety) is an *r-form in n variables.*

In the expression α above, we notice the absence of terms in $dz\,dy$, $dx\,dz$, $dy\,dx$, which suggests symmetry or skew-symmetry. The further absence of terms $dx\,dx$, $\cdots$ strongly suggests the latter.

We shall set up a calculus of differential forms which will have certain inner consistency properties, one of which is the rule for changing variables in a multiple integral. Our integrals are always oriented integrals, hence we never take absolute values of Jacobians.

Consider

$$\iint A(x, y)\,dx\,dy$$

with the change of variable

$$\begin{cases} x = x(u, v) \\ y = y(u, v). \end{cases}$$

We have

$$\iint A(x, y)\, dx\, dy = \iint A[x(u, v), y(u, v)] \frac{\partial(x, y)}{\partial(u, v)}\, du\, dv,$$

which leads us to write

$$dx\, dy = \frac{\partial(x, y)}{\partial(u, v)}\, du\, dv = \begin{vmatrix} \dfrac{\partial x}{\partial u} & \dfrac{\partial x}{\partial v} \\ \dfrac{\partial y}{\partial u} & \dfrac{\partial y}{\partial v} \end{vmatrix} du\, dv.$$

If we set $y = x$, the determinant has equal rows, hence vanishes. Also if we interchange x and y, the determinant changes sign. This motivates the rules

$$\begin{cases} dx\, dx = 0 \\ dy\, dx = -dx\, dy \end{cases}$$

for multiplication of differentials in our calculus.

In general, an (*exterior*) *r-form in n variables* $x^1, \cdots, x^n$ will be an expression

$$\omega = \frac{1}{r!} \sum A_{i_1, \ldots, i_r}\, dx^{i_1} \cdots dx^{i_r},$$

where the coefficients A are smooth functions of the variables and skew-symmetric in the indices.

We shall associate with each r-form ω an $(r + 1)$-form $d\omega$ called the exterior derivative of ω. Its definition will be given in such a way that validates the general Stokes' formula

$$\int_{\partial\Sigma} \omega = \int_{\Sigma} d\omega.$$

Here $\boldsymbol{\Sigma}$ is an $(r + 1)$-dimensional oriented variety and $\partial\boldsymbol{\Sigma}$ is its boundary.

A basic relation is the Poincaré Lemma:

$$d\,(d\omega) = 0.$$

In all cases this reduces to the equality of mixed second partials.

1.2. Comparison with Tensors

At the outset we can assure our readers that we shall not do away with tensors by introducing differential forms. Tensors are here to stay; in a great many situations, particularly those dealing with symmetries, tensor methods are very natural and effective. However, in many other situations the use of the exterior calculus, often combined with the method of moving frames of É. Cartan, leads to decisive results in a way which is very difficult with tensors alone. Sometimes a combination of techniques is in order. We list several points of contrast.

(a) Tensor analysis *per se* seems to consist only of techniques for calculations with indexed quantities. It lacks a body of substantial or deep results established once and for all within the subject and then available for application. The exterior calculus does have such a body of results.

If one takes a close look at Riemannian geometry as it is customarily developed by tensor methods one must seriously ask whether the geometric results cannot be obtained more cheaply by other machinery.

(b) In classical tensor analysis, one never knows what is the range of applicability simply because one is never told what the space is. Everything seems to work in a coordinate patch, but we know this is inadequate for most applications. For example, if a particle is constrained to move on the sphere $\mathbf{S}^2$, a single coordinate system cannot describe its position space, let alone its phase or state spaces.

This difficulty has been overcome in modern times by the theory of differentiable manifolds (varieties) which we discuss in Chapter V.

(c) Tensor fields do not behave themselves under mappings. For example, given a contravariant vector field a^i on x-space and a mapping ϕ on x-space to y-space, there is no naturally induced field on the y-space. [Try the map $t \longrightarrow (t^2, t^3)$ on $\mathbf{E}^1$ into $\mathbf{E}^2$.]

With exterior forms we have a really attractive situation in this regard. If

$$\phi\colon \quad \mathbf{M} \longrightarrow \mathbf{N}$$

and if ω is a p-form on $\mathbf{N}$, there is naturally induced a p-form $\phi^* \omega$ on $\mathbf{M}$.

Let us illustrate this for the simplest case in which ω is a 0-form, or scalar, i.e., a real-valued function on $\mathbf{N}$. Here $\phi^* \omega = \omega \circ \phi$, the composition of the mapping ϕ followed by ω.

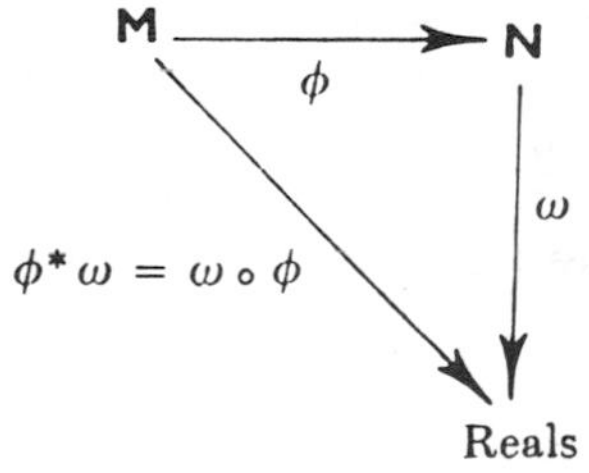

(d) In tensor calculations the maze of indices often makes one lose sight of the very great differences between various types of quantities which can be represented by tensors, for example, vectors tangent to a space, mappings between such vectors, geometric structures on the tangent spaces.

(e) It is often quite difficult using tensor methods to discover the deeper invariants in geometric and physical situations, even the local ones. Using exterior forms, they seem to come naturally according to these principles:

(i) All local geometric relations arise one way or another from the equality of mixed partials, i.e., Poincaré's Lemma.

(ii) Local invariants themselves usually appear as the result of applying exterior differentiation to everything in sight.

(iii) Global relations arise from integration by parts, i.e., Stokes' theorem.

(iv) Existence problems which are not genuine partial differential equations (boundary value or Cauchy problems) generally are of the type of Frobenius–Cartan–Kähler system of exterior differential forms and can be reduced thereby to systems of ordinary equations.

(f) In studying geometry by tensor methods, one is invariably restricted to the *natural frames* associated with a local coordinate system. Let us consider a Riemannian geometry as a case in point. This consists of a manifold in which a Euclidean geometry has been imposed in each of the tangent spaces. A natural frame leads to an oblique coordinate system in each tangent space. Now who in his right mind would study Euclidean geometry with oblique coordinates? Of course the orthonormal coordinate systems are the natural ones for Euclidean geometry, so they must be the correct ones for the much harder Riemannian geometry. We are led to introduce moving frames, a method which goes hand-in-glove with exterior forms.

We conclude the case by stating our opinion, that exterior calculus is here to stay, that it will gradually replace tensor methods in numerous situations where it is the more natural tool, that it will find more and more applications because of its inner simplicity, body of substantial results begging for further use, and because it simply is there wherever integrals occur. There is generally a time lag of some fifty years between mathematical theories and their applications. The mathematicians H. Poincaré, É. Goursat, and É. Cartan developed the exterior calculus in the early part of this century; in the last twenty years it has greatly contributed to the rebirth of differential geometry, now part of the mathematical main stream. Physicists are beginning to realize its usefulness; perhaps it will soon make its way into engineering.

II

Exterior Algebra

2.1. The Space of p-Vectors

Notation:

$\mathbf{R}$ = field of real numbers $a, b, c, \cdots$.

$\mathbf{L}$ = an n-dimensional vector space over $\mathbf{R}$ with elements $\alpha, \beta, \cdots$.

For each $p = 0, 1, 2, \cdots, n$ we shall construct a new vector space

$$\bigwedge^p \mathbf{L}$$

over $\mathbf{R}$, called *the space of p-vectors on* $\mathbf{L}$. We begin with

$$\bigwedge^0 \mathbf{L} = \mathbf{R}, \qquad \bigwedge^1 \mathbf{L} = \mathbf{L}.$$

Next we shall work out $\bigwedge^2 \mathbf{L}$ in some detail. This space consists of all sums

$$\sum a_i(\alpha_i \wedge \beta_i)$$

subject only to these constraints, or reduction rules, and no others:

$$\left\{\begin{aligned} (a_1\alpha_1 + a_2\alpha_2) \wedge \beta - a_1(\alpha_1 \wedge \beta) - a_2(\alpha_2 \wedge \beta) &= 0, \\ \alpha \wedge (b_1\beta_1 + b_2\beta_2) - b_1(\alpha \wedge \beta_1) - b_2(\alpha \wedge \beta_2) &= 0, \\ \alpha \wedge \alpha &= 0, \\ \alpha \wedge \beta + \beta \wedge \alpha &= 0. \end{aligned}\right.$$

Here α, β, etc., are vectors in $\mathbf{L}$ and a, b, etc., are real numbers; $\alpha \wedge \beta$ is called the *exterior product* of the vectors α and β. If α and β are dependent, say $\beta = c\alpha$, then

$$\alpha \wedge \beta = \alpha \wedge (c\alpha) = c(\alpha \wedge \alpha) = c \cdot 0 = 0$$

according to our reductions. Otherwise $\alpha \wedge \beta \neq 0$.

Suppose $\sigma^1, \cdots, \sigma^n$ is a basis of $\mathbf{L}$. Then

$$\alpha = \sum a_i\sigma^i, \qquad \beta = \sum b_j\sigma^j,$$

$$\alpha \wedge \beta = \left(\sum a_i\sigma^i\right) \wedge \left(\sum b_j\sigma^j\right) = \sum a_i b_j(\sigma^i \wedge \sigma^j).$$

We rearrange this as follows. Each term $\sigma^i \wedge \sigma^i = 0$ and each $\sigma^j \wedge \sigma^i = -\sigma^i \wedge \sigma^j$ for $i < j$. Hence

$$\alpha \wedge \beta = \sum_{i<j} (a_i b_j - a_j b_i)\sigma^i \wedge \sigma^j.$$

The typical element of $\bigwedge^2 \mathbf{L}$ is a linear combination of such exterior products, hence the 2-vectors

$$\sigma^i \wedge \sigma^j, \qquad 1 \leqq i < j \leqq n,$$

form a basis of $\bigwedge^2 \mathbf{L}$. We conclude

$$\dim \bigwedge^2 \mathbf{L} = \frac{n(n-1)}{2} = \binom{n}{2}.$$

In general, we form $\bigwedge^p \mathbf{L}$ $(2 \leqq p \leqq n)$ by the same idea. It consists of all formal sums (*p-vectors*, or vectors of degree p)

$$\sum a(\alpha_1 \wedge \cdots \wedge \alpha_p)$$

subject only to these constraints:

(i) $(a\alpha + b\beta) \wedge \alpha_2 \wedge \cdots \wedge \alpha_p$
$$= a(\alpha \wedge \alpha_2 \wedge \cdots \wedge \alpha_p) + b(\beta \wedge \alpha_2 \wedge \cdots \wedge \alpha_p),$$
and the same if any α_i is replaced by a linear combination.

(ii) $\alpha_1 \wedge \cdots \wedge \alpha_p = 0$ if for some pair of indices $i \neq j$, $\alpha_i = \alpha_j$.

(iii) $\alpha_1 \wedge \cdots \wedge \alpha_p$ changes sign if any two α_i are interchanged.

It follows easily from (i) that $\alpha_1 \wedge \cdots \wedge \alpha_p$ is linear in each variable; we may replace any variable by a linear combination of any number (not just two) of other vectors and compute the value by distributing, for example

$$\alpha \wedge (b_1\beta_1 + b_2\beta_2 + b_3\beta_3) \wedge \gamma \wedge \delta$$
$$= b_1(\alpha \wedge \beta_1 \wedge \gamma \wedge \delta) + b_2(\alpha \wedge \beta_2 \wedge \gamma \wedge \delta) + b_3(\alpha \wedge \beta_3 \wedge \gamma \wedge \delta).$$

It follows from (iii) that if π is any permutation of the $\{1, 2, \cdots, p\}$, then

$$\alpha_{\pi(1)} \wedge \cdots \wedge \alpha_{\pi(p)} = (\operatorname{sgn} \pi)\, \alpha_1 \wedge \cdots \wedge \alpha_p .$$

Exactly as in the case $p = 2$, we can show that if

$$\sigma^1, \cdots, \sigma^n$$

is a basis of $\mathbf{L}$, then a basis of $\bigwedge^p \mathbf{L}$ is made up as follows: for each set of indices

$$H = \{h_1, h_2, \cdots, h_p\}, \qquad 1 \leqq h_1 < h_2 < \cdots < h_p \leqq n,$$

we set

$$\sigma^H = \sigma^{h_1} \wedge \cdots \wedge \sigma^{h_p}.$$

Then the totality of σ^H is a basis of $\bigwedge^p \mathbf{L}$. We conclude that

$$\dim \bigwedge^p \mathbf{L} = \binom{n}{p},$$

the number of combinations of n things taken p at a time. In particular

$$\dim \bigwedge^n \mathbf{L} = 1.$$

If λ is in $\bigwedge^p \mathbf{L}$, then

$$\lambda = \sum_H a_H \sigma^H,$$

summed over all of these ordered sets H. One can also sum over all p-tuples of indices by introducing skew-symmetric coefficients:

$$\lambda = \frac{1}{p!} \sum_{h_1, \cdots, h_p} b_{h_1, \cdots, h_p} \sigma^{h_1} \wedge \cdots \wedge \sigma^{h_p}$$

where the $b_{h_1 \cdots h_p}$ is a skew-symmetric tensor and

$$b_{h_1 \cdots h_p} = a_H \qquad \text{for} \qquad H = \{h_1, \cdots, h_p\}, \qquad h_1 < h_2 < \cdots < h_p .$$

This skew-symmetric representation is often quite useful.

Let us note why we do not define $\bigwedge^p \mathbf{L}$ for $p > n$. (Sometimes it is convenient to simply set $\bigwedge^p \mathbf{L} = 0$ for $p > n$.) We express each α in a product $\alpha_1 \wedge \cdots \wedge \alpha_p$ as a linear combination of the basis vectors $\sigma^1, \cdots, \sigma^n$ and completely distribute according to Rule (i). This leads to

$$\alpha_1 \wedge \cdots \wedge \alpha_p = \sum a_{h_1 \cdots h_p} \sigma^{h_1} \wedge \cdots \wedge \sigma^{h_p}.$$

Each term $\sigma^{h_1} \wedge \cdots \wedge \sigma^{h_p}$ is a product of $p > n$ vectors taken from the set $\sigma^1, \cdots, \sigma^n$ so there must be a repetition; by Rule (ii) it vanishes. We are left with $\alpha_1 \wedge \cdots \wedge \alpha_p = 0$ as the only possibility.

We close with a very important property of the spaces $\bigwedge^p \mathbf{L}$.

In order to define a linear mapping f on $\bigwedge^p \mathbf{L}$ it suffices to present a function g of p variables on $\mathbf{L}$ such that (i) g is linear in each variable separately, (ii) g is alternating in the sense that g vanishes when two of its variables are equal and g changes sign when two of its variables are interchanged. Then

$$f(\alpha_1 \wedge \cdots \wedge \alpha_p) = g(\alpha_1, \cdots, \alpha_p)$$

defines f on the generators of $\bigwedge^p \mathbf{L}$.

It can be shown that this property provides an axiomatic characterization of $\bigwedge^p \mathbf{L}$. In the next section we apply this property to define the determinant of a linear transformation.

2.2. Determinants

As above $\mathbf{L}$ is a fixed linear space of dimension n. Let A be a linear transformation on $\mathbf{L}$ into itself. We define a function $g = g_A$ of n variables on $\mathbf{L}$ as follows:

$$g_A(\alpha_1, \cdots, \alpha_n) = A\alpha_1 \wedge \cdots \wedge A\alpha_n,$$

$$g_A\colon \ \bigtimes^n \mathbf{L} \longrightarrow \bigwedge^n \mathbf{L}$$

where $\bigtimes^n \mathbf{L}$ denotes the cartesian product. Since g is multilinear and alternating, there is a linear functional $f = f_A$,

$$f_A\colon\ \bigwedge^n \mathbf{L} \longrightarrow \bigwedge^n \mathbf{L}$$

satisfying

$$f_A(\alpha_1 \wedge \cdots \wedge \alpha_n) = g_A(\alpha_1, \cdots, \alpha_n) = A\alpha_1 \wedge \cdots \wedge A\alpha_n.$$

But $\bigwedge^n \mathbf{L}$ is one-dimensional so the only linear transformation on this space is multiplication by a scalar. We denote the particular one here by $|A|$ and have

$$A\alpha_1 \wedge \cdots \wedge A\alpha_n = |A|(\alpha_1 \wedge \cdots \wedge \alpha_n).$$

This serves to define the *determinant* $|A|$ of A. We must not fail to note that this definition is completely independent of a matrix representation of A.

We observe next

$$\begin{aligned}|AB|(\alpha_1 \wedge \cdots \wedge \alpha_n) &= (AB\alpha_1) \wedge \cdots \wedge (AB\alpha_n)\\ &= |A|(B\alpha_1 \wedge \cdots \wedge B\alpha_n)\\ &= |A|\cdot|B|(\alpha_1 \wedge \cdots \wedge \alpha_n),\end{aligned}$$

hence

$$|AB| = |A|\cdot|B|.$$

We can relate this to the determinant of a matrix as follows. Let $\sigma^1, \cdots, \sigma^n$ be a basis of $\mathbf{L}$ and $\|a_{ij}\|$ an $n \times n$ matrix. Set

$$\alpha_i = \sum a_{ij}\sigma^j.$$

Then

$$\alpha_1 \wedge \cdots \wedge \alpha_n = |a_{ij}|\sigma^1 \wedge \cdots \wedge \sigma^n.$$

In particular, if one obtains the matrix representation of A with respect to the basis (σ^i) by

$$A\sigma^i = \sum a^i{}_j\sigma^j,$$

then $$A\sigma^1 \wedge \cdots \wedge A\sigma^n = |a^i{}_j|\sigma^1 \wedge \cdots \wedge \sigma^n, \qquad |A| = |a^i{}_j|.$$

2.3. Exterior Products

We now observe that our spaces $\bigwedge^p \mathbf{L}$ have a built-in multiplication process called *exterior multiplication* and denoted by $\wedge$ for obvious reasons. We multiply a p-vector μ by a q-vector ν to obtain a $(p + q)$-vector $\mu \wedge \nu$ (which is 0 by definition if $p + q > n$):

$$\wedge\colon\ (\bigwedge^p \mathbf{L}) \times (\bigwedge^q \mathbf{L}) \longrightarrow \bigwedge^{p+q} \mathbf{L}.$$

It suffices to define $\wedge$ on generators and use the basic principle at the end of Section 1 to extend it to all p- and q-vectors:

$$(\alpha_1 \wedge \cdots \wedge \alpha_p) \wedge (\beta_1 \wedge \cdots \wedge \beta_q) = \alpha_1 \wedge \cdots \wedge \alpha_p \wedge \beta_1 \wedge \cdots \wedge \beta_q.$$

The basic properties of this exterior product are

(1) $\lambda \wedge \mu$ is distributive,

(2) $\lambda \wedge (\mu \wedge \nu) = (\lambda \wedge \mu) \wedge \nu$, the associative law

(3) $\mu \wedge \lambda = (-1)^{pq} \lambda \wedge \mu$.

Property (3) simply says that any two vectors of odd degrees anticommute, otherwise vectors commute. The following will illustrate why this is the case:

$$\begin{aligned}(\alpha_1 \wedge \alpha_2 \wedge \alpha_3) \wedge \beta &= -(\alpha_1 \wedge \alpha_2 \wedge \beta \wedge \alpha_3)\\ &= (-1)^2(\alpha_1 \wedge \beta \wedge \alpha_2 \wedge \alpha_3)\\ &= (-1)^3 \beta \wedge (\alpha_1 \wedge \alpha_2 \wedge \alpha_3),\\ (\alpha_1 \wedge \alpha_2 \wedge \alpha_3) \wedge (\beta_1 \wedge \beta_2) &= (-1)^3 \beta_1 \wedge (\alpha_1 \wedge \alpha_2 \wedge \alpha_3) \wedge \beta_2\\ &= (-1)^3(-1)^3(\beta_1 \wedge \beta_2) \wedge (\alpha_1 \wedge \alpha_2 \wedge \alpha_3)\\ &= (-1)^{3\cdot 2}(\beta_1 \wedge \beta_2) \wedge (\alpha_1 \wedge \alpha_2 \wedge \alpha_3).\end{aligned}$$

Examples. We take for **L** the linear space based on the differentials $dx, dy, \cdots$ and, as is customary, omit the exterior multiplication sign $\wedge$ between dx's. Thus $dx\,dy$ denotes $dx \wedge dy$.

1. $(A\,dx + B\,dy + C\,dz) \wedge (E\,dx + F\,dy + G\,dz)$

$$= (BG - CF)\,dy\,dz + (CE - AG)\,dz\,dx + (AF - BE)\,dx\,dy,$$

illustrating the vector-, or cross-product of two ordinary vectors.

2. $(A\,dx + B\,dy + C\,dz) \wedge (P\,dy\,dz + Q\,dz\,dx + R\,dx\,dy)$

$$= (AP + BQ + CR)\,dx\,dy\,dz,$$

illustrating the dot-, or inner-product of vector algebra.

3. Let α be any form of odd degree. Then

$$\alpha^2 = \alpha \wedge \alpha = 0.$$

For if α and β are of odd degree p, then

$$\beta \wedge \alpha = -\alpha \wedge \beta.$$

We set $\beta = \alpha$ to have

$$\alpha \wedge \alpha = -\alpha \wedge \alpha, \qquad 2(\alpha \wedge \alpha) = 0, \qquad \alpha \wedge \alpha = 0.$$

4. Here we take

$$\omega = dp_1\,dq^1 + \cdots + dp_n\,dq^n,$$

a form arising in mechanics. The two-forms $dp_i\,dq^i$ all commute, hence

$$\begin{aligned}\omega^n &= (n!)\,dp_1\,dq^1\,dp_2\,dq^2 \cdots dp_n\,dq^n\\ &= (-1)^{n(n-1)/2}(n!)\,dp_1 \cdots dp_n\,dq^1 \cdots dq^n.\end{aligned}$$

The product $dp_1 \cdots dq^n$ is called *phase-density*. We shall discuss this further in Chapter X.

We apply the exterior product to obtain the Laplace expansion of a determinant by complementary minors.

Let $\|a_{ij}\|$ be an $n \times n$ matrix. For $H = \{h_1, \cdots, h_p\}$, set

$$b_H = \begin{vmatrix} a_{1,h_1} & \cdots & a_{1,h_p} \\ \vdots & & \\ a_{p,h_1} & \cdots & a_{p,h_p} \end{vmatrix}.$$

Set $p + q = n$. For $K = \{k_1, \cdots, k_q\}$, set

$$c_K = \begin{vmatrix} a_{p+1,k_1} & \cdots & a_{p+1,k_q} \\ \vdots & & \\ a_{n,k_1} & \cdots & a_{n,k_q} \end{vmatrix}.$$

Thus if $K = H'$, the complementary set of indices to H (always arranged in natural order), then b_H and c_K are complementary minors of $\|a_{ij}\|$.

Now set

$$\alpha_i = \sum a_{ij}\sigma^j$$

where (σ^j) is a basis of **L**. We easily see that

$$\alpha_1 \wedge \cdots \wedge \alpha_p = \sum b_H \sigma^H,$$
$$\alpha_{p+1} \wedge \cdots \wedge \alpha_n = \sum c_K \sigma^K,$$

hence

$$\alpha_1 \wedge \cdots \wedge \alpha_n = (\alpha_1 \wedge \cdots \wedge \alpha_p) \wedge (\alpha_{p+1} \wedge \cdots \wedge \alpha_n)$$
$$= \sum b_H c_K \sigma^H \wedge \sigma^K.$$

But

$$\alpha_1 \wedge \cdots \wedge \alpha_n = |a_{ij}|(\sigma^1 \wedge \cdots \wedge \sigma^n)$$

and

$$\sigma^H \wedge \sigma^K = \begin{cases} 0 & \text{if } K \neq H' \\ \varepsilon^{H,H'}(\sigma^1 \wedge \cdots \wedge \sigma^n) & \text{if } K = H', \end{cases}$$

hence

$$|a_{ij}| = \sum_H \varepsilon^{H,H'} b_H c_{H'}.$$

If $H = \{h_1, \cdots, h_p\}$, $H' = \{k_1, \cdots, k_q\}$, then

$$\varepsilon^{H,H'} = \operatorname{sgn}\begin{pmatrix} 1 & 2 & \cdots & n \\ h_1 & h_2 & \cdots & k_q \end{pmatrix}.$$

2.4. Linear Transformations

In this section we deal with two linear spaces **M** and **N** with

$$\dim \mathbf{M} = m, \qquad \dim \mathbf{N} = n.$$

Let us agree that when we need bases, $\sigma^1, \cdots, \sigma^m$ will denote a basis of $\mathbf{M}$ and $\tau^1, \cdots, \tau^n$ a basis of $\mathbf{N}$.

Let A be a linear transformation,

$$A\colon\quad \mathbf{M} \longrightarrow \mathbf{N}.$$

The mapping

$$(\alpha_1, \cdots, \alpha_p) \longrightarrow A\alpha_1 \wedge \cdots \wedge A\alpha_p$$

sends

$$\textstyle\bigtimes^p \mathbf{M} \longrightarrow \bigwedge^p \mathbf{N}.$$

It is alternating multilinear, hence defines a linear transformation, denoted $\wedge^p A$ on $\bigwedge^p \mathbf{M}$ to $\bigwedge^p \mathbf{N}$. This *exterior* pth power of A is defined on generators by

$$(\wedge^p A)(\alpha_1 \wedge \cdots \wedge \alpha_p) = A\alpha_1 \wedge \cdots \wedge A\alpha_p.$$

Suppose A is represented by the $m \times n$ matrix $\|a^i{}_j\|$ according to

$$A\sigma^i = \sum a^i{}_j \tau^j.$$

The σ^H and τ^K form bases of $\bigwedge^p \mathbf{M}$ and $\bigwedge^p \mathbf{N}$, respectively, where H and K are ordered sets of p indices. We have

$$\begin{aligned}(\wedge^p A)\sigma^H &= A\sigma^{h_1} \wedge \cdots \wedge A\sigma^{h_p}\\ &= \sum a^{h_1}{}_{k_1} \cdots a^{h_p}{}_{k_p}\, \tau^{k_1} \wedge \cdots \wedge \tau^{k_p}\\ &= \sum a^H{}_K\, \tau^K.\end{aligned}$$

Hence $\wedge^p A$ is represented by the matrix

$$\|a^H{}_K\|$$

of all $p \times p$ minors of $\|a^i{}_j\|$. This is sometimes called the pth *compound* of $\|a^i{}_j\|$.

Suppose one has three spaces $\mathbf{L}$, $\mathbf{M}$, $\mathbf{N}$ and this situation:

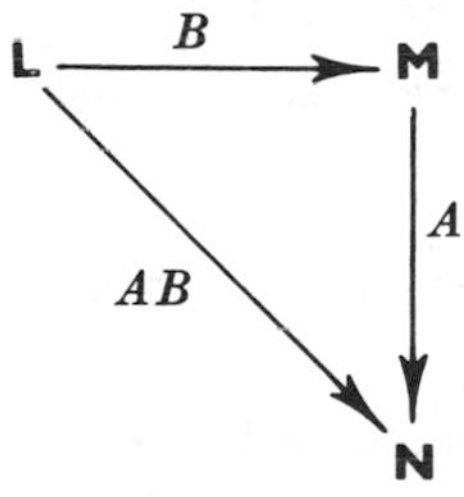

We compute $\wedge^p (AB)$:

$$\begin{aligned}\wedge^p (AB)(\alpha_1 \wedge \cdots \wedge \alpha_p) &= (AB\alpha_1) \wedge \cdots \wedge (AB\alpha_p)\\ &= (\wedge^p A)[(B\alpha_1) \wedge \cdots \wedge (B\alpha_p)]\\ &= (\wedge^p A)[(\wedge^p B)(\alpha_1 \wedge \cdots \wedge \alpha_p)]\\ &= [(\wedge^p A)(\wedge^p B)](\alpha_1 \wedge \cdots \wedge \alpha_p),\end{aligned}$$

hence

$$\wedge^p (AB) = (\wedge^p A)(\wedge^p B).$$

It follows that the pth compound of the product of two matrices is the product of their pth compounds, a nontrivial result.

We must consider one other matter. Again let A: $\mathbf{M} \longrightarrow \mathbf{N}$. Suppose ω is in $\bigwedge^p \mathbf{M}$ and η is in $\bigwedge^q \mathbf{M}$. Then

$$(\wedge^{p+q} A)(\omega \wedge \eta) = (\wedge^p A)(\omega) \wedge (\wedge^q A)(\eta).$$

For if we take monomials, $\omega = \alpha_1 \wedge \cdots \wedge \alpha_p$, $\eta = \beta_1 \wedge \cdots \wedge \beta_q$, then

$$\begin{aligned}(\wedge^{p+q} A)(\omega \wedge \eta) &= (\wedge^{p+q} A)(\alpha_1 \wedge \cdots \wedge \alpha_p \wedge \beta_1 \wedge \cdots \wedge \beta_q)\\ &= A\alpha_1 \wedge \cdots \wedge A\beta_q\\ &= (A\alpha_1 \wedge \cdots \wedge A\alpha_p) \wedge (A\beta_1 \wedge \cdots \wedge A\beta_q)\\ &= (\wedge^p A)(\omega) \wedge (\wedge^q A)(\eta).\end{aligned}$$

2.5. Inner Product Spaces

In the remainder of the chapter we shall study a space $\mathbf{L}$ which has an inner product (α, β). This is a real-valued function on $\mathbf{L} \times \mathbf{L}$ which is

(i) Linear in each variable,
(ii) Symmetric: $(\alpha, \beta) = (\beta, \alpha)$,
(iii) Nondegenerate: if for fixed α, $(\alpha, \beta) = 0$ for all β, then $\alpha = 0$.

Example 1. The *Euclidean* inner product on $\mathbf{E}^n$ is given by

$$\alpha = (a_1, \cdots, a_n), \qquad \beta = (b_1, \cdots, b_n),$$
$$(\alpha, \beta) = a_1 b_1 + \cdots + a_n b_n.$$

Example 2. The *Lorentz* inner product in four-space:

$$\alpha = (a_1, \cdots, a_4), \qquad \beta = (b_1, \cdots, b_4),$$
$$(\alpha, \beta) = a_1 b_1 + a_2 b_2 + a_3 b_3 - c^2 a_4 b_4$$

where c is the speed of light.

Condition (iii) is equivalent to the following. If $\sigma^1, \cdots, \sigma^n$ is a basis of $\mathbf{L}$, then

$$|(\sigma^i, \sigma^j)| \neq 0.$$

(The left-hand member is the Gram determinant, or *Grammian.*) For this determinant vanishes if and only if there is a nontrivial solution $(a_1, \cdots, a_n)$ of the homogeneous system

$$\sum a_i(\sigma^i, \sigma^j) = 0.$$

But this is the same as having the vector

$$\alpha = \sum a_i \sigma^i$$

satisfy the relation $(\alpha, \beta) = 0$ for all β.

An *orthonormal basis* of $\mathbf{L}$ consists of a basis $\sigma^1, \cdots, \sigma^n$ such that

$$(\sigma^i, \sigma^j) = \pm \delta^{ij}.$$

If there are r plus signs and s minus signs, then $r + s = n$, and $t = r - s$ is the *signature* of the inner product. It does not depend on the choice of basis.

It is a basic fact that *each inner product space* $\mathbf{L}$ *has an orthonormal basis.* This is proved in several steps.

1. If $\dim \mathbf{L} > 0$, there is a vector σ in $\mathbf{L}$ such that $(\sigma, \sigma) \neq 0$. For if $(\alpha, \alpha) = 0$ for all α, then

$$\begin{aligned} 0 = (\alpha + \beta, \alpha + \beta) &= (\alpha, \alpha) + 2(\alpha, \beta) + (\beta, \beta) \\ &= 2(\alpha, \beta), \qquad (\alpha, \beta) = 0 \qquad \text{for all } \alpha, \beta, \end{aligned}$$

a contradiction to nondegeneracy.

2. Pick a *maximal* sequence $\sigma^1, \cdots, \sigma^r$ of vectors satisfying

$$(\sigma^i, \sigma^j) = \pm \delta^{ij}.$$

Let $\mathbf{M}$ be the subspace of $\mathbf{L}$ these vectors span. Then $\dim \mathbf{M} = r$. [The σ^i are independent since $\sum a_i \sigma^i = 0$ implies $\sum a_i(\sigma^i, \sigma^j) = 0$, $\pm a_j = 0$.] We suppose $r < n$.

3. Let $\mathbf{N}$ be the orthogonal complement of $\mathbf{M}$, i.e., $\mathbf{N}$ is the space of all vectors β such that $(\alpha, \beta) = 0$ for all α in $\mathbf{M}$. Since $\mathbf{N}$ is determined by the r relations $(\sigma^i, \beta) = 0$, $\dim \mathbf{N} \geqq n - r$. But obviously $\mathbf{M} \cap \mathbf{N} = 0$ (i.e., the only vector common to $\mathbf{M}$ and $\mathbf{N}$ is 0), hence $\dim \mathbf{N} = n - r$, $\mathbf{M}$ and $\mathbf{N}$ together span $\mathbf{L}$, $\mathbf{M} + \mathbf{N} = \mathbf{L}$.

4. $\mathbf{N}$ itself is an inner product space relative to the inner product of $\mathbf{L}$. Only the property (iii) of nondegeneracy must be checked. Suppose β is in $\mathbf{N}$ and $(\gamma, \beta) = 0$ for all γ in $\mathbf{N}$. But $(\alpha, \beta) = 0$ for all α in $\mathbf{M}$, hence $(\alpha, \beta) = 0$ for all α in $\mathbf{L}$ since $\mathbf{M}$ and $\mathbf{N}$ together span $\mathbf{L}$. Hence $\beta = 0$.

5. By (1), there is a vector α in $\mathbf{N}$ such that $(\alpha, \alpha) \neq 0$. We set

$$\sigma^{r+1} = \alpha / |(\alpha, \alpha)|^{1/2}$$

and see that we have constructed a sequence $\sigma^1, \cdots, \sigma^{r+1}$ longer than a maximal one. Since this is impossible we conclude that we must have had $r = n$ in the first place, which completes the proof.

There is another basic property of inner product spaces which we shall need below.

Let f be a linear functional on $\mathbf{L}$. *Then there is a unique vector β in* $\mathbf{L}$ *such that*

$$f(\alpha) = (\alpha, \beta).$$

This is easily established by taking an orthonormal basis $\sigma^1, \cdots, \sigma^n$. We set $b_i = f(\sigma^i)$ and for β simply take

$$\beta = \sum \pm b_j \sigma^j = \sum (\sigma^j, \sigma^j) b_j \sigma^j.$$

For then

$$(\sigma^i, \beta) = \sum_j (\sigma^j, \sigma^j) b_j (\sigma^i, \sigma^j) = b_j = f(\sigma^i).$$

2.6. Inner Products of p-Vectors

Again we start with an n-dimensional vector space $\mathbf{L}$ with an inner product (α, β). We shall define an induced inner product on each of the spaces $\bigwedge^p \mathbf{L}$. We set

$$(\lambda, \mu) = |(\alpha_i, \beta_j)|$$

for $\lambda = \alpha_1 \wedge \cdots \wedge \alpha_p$, $\mu = \beta_1 \wedge \cdots \wedge \beta_p$. This definition works because the determinant on the right is an alternating multilinear function of the α's, ditto the β's. This means the formula defines a scalar-valued function on $(\bigwedge^p \mathbf{L}) \times (\bigwedge^p \mathbf{L})$ which is linear in each variable. Next $(\mu, \lambda) = (\lambda, \mu)$ because interchanging the rows and columns of a matrix (transposing) does not change its determinant.

The nondegeneracy of this inner product is most easily seen by computing with respect to an orthonormal basis $\sigma^1, \cdots, \sigma^n$ of $\mathbf{L}$. As usual the σ^H, $H = \{h_1 < h_2 < \cdots < h_p\}$, form a basis of $\bigwedge^p \mathbf{L}$. We have

$$(\sigma^H, \sigma^K) = |(\sigma^{h_i}, \sigma^{k_j})|.$$

If $H \neq K$, this is zero since the determinant has a row (also a column) of zeros. If $H = K$, all but the diagonal elements vanish and these are ± 1, hence

$$(\sigma^H, \sigma^K) = \pm \delta^{H,K}$$

In other words, the σ^H form an orthonormal basis of $\bigwedge^p \mathbf{L}$, nondegeneracy follows free of charge.

In particular $\sigma = \sigma^1 \wedge \cdots \wedge \sigma^n$ is an orthonormal basis of $\bigwedge^n \mathbf{L}$ and

$$(\sigma, \sigma) = (\sigma^1, \sigma^1) \cdots (\sigma^n, \sigma^n) = (-1)^{(n-t)/2}$$

where t is the signature of $\mathbf{L}$.

For another example, set

$$\alpha^i = \sigma^1 \wedge \cdots \wedge \sigma^{i-1} \wedge \sigma^{i+1} \wedge \cdots \wedge \sigma^n,$$

forming a basis of $\bigwedge^{n-1} \mathbf{L}$. Clearly

$$(\alpha^i, \alpha^i) = (\sigma, \sigma)/(\sigma^i, \sigma^i) = (\sigma, \sigma)(\sigma^i, \sigma^i),$$

hence

$$\begin{aligned}(\sum a_i \alpha^i, \sum b_j \alpha^j) &= (\sigma, \sigma) \sum (\sigma^i, \sigma^i) a_i b_i \\ &= (\sigma, \sigma)(\sum a_i \sigma^i, \sum b_j \sigma^j).\end{aligned}$$

2.7. The Star Operator

Again let **L** have inner product (α, β). We shall take a definite orientation of **L** which will remain fixed. (This simply means we take one basis for **L** and only consider other bases which are expressed in terms of this one by a matrix with positive determinant. The space **L** has two orientations and we take one of them.) We only use bases coherent to the orientation.

We shall define an operation $*$, called the (*Hodge*) *star operator*. This will be a linear transformation on $\bigwedge^p \mathbf{L}$ onto $\bigwedge^{n-p} \mathbf{L}$. This operator depends, of course, on the inner product and also depends on the orientation. Reversing orientation will change its sign.

We note that the orientation of **L** determines a definite orthonormal basis σ of $\bigwedge^n \mathbf{L}$.

Now fix λ in $\bigwedge^p \mathbf{L}$. The mapping

$$\mu \longrightarrow \lambda \wedge \mu$$

is a linear transformation on $\bigwedge^{n-p} \mathbf{L}$ into the one-dimensional space $\bigwedge^n \mathbf{L}$. We may write

$$\lambda \wedge \mu = f_\lambda(\mu)\,\sigma$$

where f_λ is a linear functional on $\bigwedge^{n-p} \mathbf{L}$. By our result at the end of Section 2.5, there is a unique $(n-p)$-vector, which we denote $*\lambda$ to indicate its dependence on λ, such that

$$\lambda \wedge \mu = (*\lambda, \mu)\sigma.$$

This equation defines the $*$ map which is evidently linear on $\bigwedge^p \mathbf{L}$ into $\bigwedge^{n-p} \mathbf{L}$.

In order to compute $*\lambda$ for generators of $\bigwedge^p \mathbf{L}$, in view of the linearity, it is enough to compute $*\lambda$ where $\lambda = \sigma^1 \wedge \cdots \wedge \sigma^p$ and where $\sigma^1, \cdots, \sigma^n$ is an orthonormal basis. Let K run over sets of $q = n - p$ indices. Then

$$\lambda \wedge \sigma^K = (*\lambda, \sigma^K)\sigma.$$

The left-hand side vanishes unless $K = \{p + 1, p + 2, \cdots, n\}$, hence

$$*\lambda = c\sigma^{p+1} \wedge \cdots \wedge \sigma^n$$

and the constant c is determined taking $K = \{p + 1, \cdots, n\}$:

$$\sigma = \lambda \wedge \sigma^K = c(\sigma^K, \sigma^K)\sigma,$$

$$c = (\sigma^K, \sigma^K) = \pm 1,$$

$$*\lambda = (\sigma^K, \sigma^K)\sigma^K.$$

For definiteness, set $H = \{1, \cdots, p\}$, $K = \{p + 1, \cdots, n\}$. We have proved

$$*\sigma^H = (\sigma^K, \sigma^K)\sigma^K.$$

Since $\sigma^K \wedge \sigma^H = (-1)^{p(n-p)}\sigma^H \wedge \sigma^K$, we deduce, taking orientation into account,

$$*\sigma^K = (-1)^{p(n-p)}(\sigma^H, \sigma^H)\sigma^H,$$

hence

$$*(*\sigma^H) = (-1)^{p(n-p)}(\sigma^H, \sigma^H)(\sigma^K, \sigma^K)\sigma^H,$$

$$\begin{aligned} *(*\sigma^H) &= (-1)^{p(n-p)}(\sigma, \sigma)\sigma^H \\ &= (-1)^{p(n-p)+(n-t)/2}\sigma^H. \end{aligned}$$

where t is the signature.

It follows that *if α is any p-vector, then*

$$**\alpha = (-1)^{p(n-p)+(n-t)/2}\alpha.$$

Another consequence of these formulas is this result.

If α, β are p-vectors, then

$$\alpha \wedge *\beta = \beta \wedge *\alpha = (-1)^{(n-t)/2}(\alpha, \beta)\sigma.$$

For when $\beta = \sigma^H$ as above, the only generator $\alpha = \sigma^J$ for which both sides do not vanish is $\alpha = \sigma^H$, and then

$$\begin{aligned} \alpha \wedge *\beta = \sigma^H \wedge (\sigma^K, \sigma^K)\sigma^K &= (\sigma^K, \sigma^K)\sigma \\ &= (\sigma^H, \sigma^H)(-1)^{(n-t)/2}\sigma \\ &= (-1)^{(n-t)/2}(\alpha, \beta)\sigma. \end{aligned}$$

Example 1. We take 4-space with coordinates so normalized that dx^1, dx^2, dx^3, dt is an orthonormal basis with $(dx^i, dx^i) = 1$, $(dt, dt) = -1$. We have $n = 4$, $t = 2$, $(-1)^{(n-t)/2} = -1$. We shall study certain two-forms. For $p = 2$, $p(n-p) = 4$. Thus

$$*(dx^i\, dt) = dx^j\, dx^k$$

where (i, j, k) is cyclic order,

$$*(dx^j\, dx^k) = -dx^i\, dt.$$

Let E_i be the components of electric field strength, H_i the components of magnetic field strength (all in free space) and consider the form

$$\omega = (E_1\, dx^1 + E_2\, dx^2 + E_3\, dx^3)\, dt + (H_1\, dx^2\, dx^3 + H_2\, dx^3\, dx^1 + H_3\, dx^1\, dx^2).$$

Then

$$\begin{aligned} *\omega = -(H_1\, dx^1 + H_2\, dx^2 + H_3\, dx^3)\, dt \\ + (E_1\, dx^2\, dx^3 + E_2\, dx^3\, dx^1 + E_3\, dx^1\, dx^2). \end{aligned}$$

We shall see the use of these forms in Maxwell's equations later.

Example 2. $\mathbf{E}^3$ with the ordinary metric. If f and g are functions,

$$df = \frac{\partial f}{\partial x}dx + \frac{\partial f}{\partial y}dy + \frac{\partial f}{\partial z}dz$$

$$*df = \frac{\partial f}{\partial x}dy\,dz + \frac{\partial f}{\partial y}dz\,dx + \frac{\partial f}{\partial z}dx\,dy$$

and we have

$$df \wedge *dg = \left(\frac{\partial f}{\partial x}\frac{\partial g}{\partial x} + \frac{\partial f}{\partial y}\frac{\partial g}{\partial y} + \frac{\partial f}{\partial z}\frac{\partial g}{\partial z}\right) dx\,dy\,dz.$$

2.8. Problems

1. Let $\mathbf{L}$ be an n-dimensional space. For each p-vector $\alpha \neq 0$ we let $\mathbf{M}_\alpha$ be the subspace of $\mathbf{L}$ consisting of all vectors σ satisfying $\alpha \wedge \sigma = 0$. Prove that $\dim(\mathbf{M}_\alpha) \leqq p$. Prove also that $\dim(\mathbf{M}_\alpha) = p$ if and only if $\alpha = \sigma_1 \wedge \cdots \wedge \sigma_p$ where $\sigma_1, \cdots, \sigma_p$ are vectors in $\mathbf{L}$.

2. (Continuation) Let α be any $(n-1)$-vector. Prove that $\alpha = \sigma_1 \wedge \cdots \wedge \sigma_{n-1}$.

3. Let $\mathbf{L}$ be an n-dimensional space and α a 2-vector. Show that there is a basis $\sigma_1, \cdots, \sigma_n$ of $\mathbf{L}$ such that

$$\alpha = \sigma_1 \wedge \sigma_2 + \sigma_3 \wedge \sigma_4 + \cdots + \sigma_{2r-1} \wedge \sigma_{2r}.$$

The number $2r$, which depends only on α, is called the *rank* of the 2-vector α. Show that $\alpha^r \neq 0$, $\alpha^{r+1} = 0$.

4. (Continuation) Let $\sigma_1, \cdots, \sigma_n$ be a basis of $\mathbf{L}$ and let $A = \|a_{ij}\|$ be a skew-symmetric matrix. Show that the rank of the matrix A coincides with the rank of the 2-vector $\alpha = \frac{1}{2}\sum a_{ij}\sigma_i \wedge \sigma_j$.

5. We are given a linear transformation

$$A\colon\ \mathbf{L} \longrightarrow \mathbf{L},$$

where $\dim \mathbf{L} = n$. Find the value of the determinant

$$|\wedge^p A|.$$

6. Let A be an $m \times n$ matrix and B an $n \times m$ matrix, where $m < n$. Prove

$$|AB| = \sum a_H b_H$$

where H runs over ordered sets,

$$H = \{h_1, h_2, \cdots, h_m\}, \qquad 1 \leqq h_1 < h_2 < \cdots < h_m \leqq n,$$

and where

$$a_H = \begin{vmatrix} a_{1h_1} & \cdots & a_{1h_m} \\ & \ddots & \\ & & a_{m,h_m} \end{vmatrix}, \qquad b_H = \begin{vmatrix} b_{h_1 1} & \cdots & b_{h_1 m} \\ & \ddots & \\ & & b_{h_m,m} \end{vmatrix}.$$

Note that the special case

$$A = \begin{pmatrix} a_1 & a_2 & a_3 \\ b_1 & b_2 & b_3 \end{pmatrix}, \qquad B = {}^tA$$

yields a well-known formula of vector algebra,

$$|\mathbf{a} \times \mathbf{b}| = |\mathbf{a}|^2|\mathbf{b}|^2 - (\mathbf{a}\cdot\mathbf{b})^2.$$

7. Let A be an $n \times n$ matrix and denote by $\operatorname{cof} A$ the matrix of cofactors of A [so that $A(\operatorname{cof} A) = (\operatorname{cof} A)A = |A|\cdot I$]. Let $b_{H,K}$ be a typical element of $\bigwedge^p (\operatorname{cof} A)$. Show that

$$b_{H,K} = \pm|A|^{p-1}a_{H',K'}$$

where H' is the set of indices complementary to H, ditto K', K, and $a_{H',K'}$ is the corresponding element of

$$\bigwedge^{n-p} A.$$

8. Express in terms of exterior algebra the formula from vector algebra

$$\boldsymbol{\alpha} \times (\boldsymbol{\beta} \times \boldsymbol{\gamma}) = (\boldsymbol{\alpha}\cdot\boldsymbol{\gamma})\boldsymbol{\beta} - (\boldsymbol{\alpha}\cdot\boldsymbol{\beta})\boldsymbol{\gamma}.$$

III

The Exterior Derivative

3.1. Differential Forms

Let P be a point in $\mathbf{E}^n$. The *one-forms* at P are the expressions

$$\sum_1^n a_i\, dx^i, \qquad a_i \text{ constants}$$

These form an n-dimensional linear space $\mathbf{L} = \mathbf{L}_P$. The *p-forms* at P are the elements of

$$\bigwedge^p \mathbf{L} = \bigwedge^p \mathbf{L}_P,$$

i.e., expressions

$$\sum a_H\, dx^{h_1} \cdots dx^{h_p}, \qquad a_H \text{ constants.}$$

Note that we are dropping the notation "$\wedge$" so that differentials dx^i juxtaposed will always be multiplied by exterior multiplication.

Now let $\mathbf{U}$ denote an (open) domain in $\mathbf{E}^n$. A *p-form on* $\mathbf{U}$ is obtained by choosing at each point P of $\mathbf{U}$ a p-form at that point, and doing this smoothly. Thus a p-form ω has the representation

$$\omega = \sum a_H(x^1, \cdots, x^n)\, dx^H,$$

where the functions $a_H(\mathbf{x})$ are smooth functions on $\mathbf{U}$, differentiable as often as we please.

The exterior algebra applies at each point of $\mathbf{U}$ and so may be interpreted on the differential forms on $\mathbf{U}$ itself. Thus if ω is a p-form and η is a q-form on $\mathbf{U}$, then $\omega \wedge \eta$ is a $(p+q)$-form on $\mathbf{U}$. (Of course $\omega \wedge \eta = 0$ if $p + q > n$.) If

$$\omega = \sum a_H\, dx^H, \qquad \eta = \sum b_K\, dx^K,$$

then

$$\omega \wedge \eta = \sum a_H b_K\, dx^H\, dx^K$$

so that the coefficients of $\omega \wedge \eta$ are again smooth functions, being polynomials in the coefficients of ω and η.

For example a one-form

$$\omega = P\,dx + Q\,dy + R\,dz$$

may be identified with an ordinary vector field (P, Q, R) in $\mathbf{E}^3$, a two-form

$$\alpha = A\, dy\, dz + B\, dz\, dx + C\, dx\, dy$$

may be identified with a polar vector field in $\mathbf{E}^3$.

3.2. Exterior Derivatives

We denote by

$$\mathbf{F}^p(\mathbf{U})$$

the totality of p-forms on $\mathbf{U}$. In particular $\mathbf{F}^0(\mathbf{U})$ is simply the set of all smooth functions on $\mathbf{U}$.

We shall now set up an operation d which takes each p-form ω to a $(p+1)$-form $d\omega$. In $\mathbf{E}^3$ it will work this way. For a 0-form f,

$$df = \frac{\partial f}{\partial x}\, dx + \frac{\partial f}{\partial y}\, dy + \frac{\partial f}{\partial z}\, dz.$$

For the one-form ω above,

$$d\omega = \left(\frac{\partial R}{\partial y} - \frac{\partial Q}{\partial z}\right) dy\, dz + \left(\frac{\partial P}{\partial z} - \frac{\partial R}{\partial x}\right) dz\, dx + \left(\frac{\partial Q}{\partial x} - \frac{\partial P}{\partial y}\right) dx\, dy,$$

while for the two-form α above,

$$d\alpha = \left(\frac{\partial A}{\partial x} + \frac{\partial B}{\partial y} + \frac{\partial C}{\partial z}\right) dx\, dy\, dz.$$

Thus the operator d subsumes the ordinary gradient, curl or rotation, and divergence.

It will turn out that d is completely independent of coordinate systems. This will be more or less clear when we axiomatize d.

We shall establish the existence and uniqueness of an operator

$$d\colon\ \mathbf{F}^p(\mathbf{U}) \longrightarrow \mathbf{F}^{p+1}(\mathbf{U})$$

such that

(i) $d(\omega + \eta) = d\omega + d\eta$
(ii) $d(\lambda \wedge \mu) = d\lambda \wedge \mu + (-1)^{(\deg \lambda)}\, \lambda \wedge d\mu$
(iii) For each ω, $d(d\omega) = 0$
(iv) For each function f,

$$df = \sum \frac{\partial f}{\partial x^i}\, dx^i.$$

Let us note the consistency of (iv) as it applies to the coordinate functions. For example x^1 is a function on $\mathbf{U}$ and $d(x^1)$ the effect of d on this function x^1 is the symbol dx^1. Thus from (iii), $d(dx^1) = 0$ once we have d.

First we prove there is only one such operation d. Suppose we are given such a d. We first show that

$$d(dx^{h_1} \cdots dx^{h_p}) = 0$$

by induction on p. We have just noted this for $p = 1$. If it is true for $p - 1$, then by (ii),

$$d[x^{h_1}(dx^{h_2} \cdots dx^{h_p})] = dx^{h_1} \cdots dx^{h_p},$$

$$d(dx^{h_1} \cdots dx^{h_p}) = d\{d(x^{h_1} dx^{h_2} \cdots dx^{h_p})\} = 0$$

by (iii). Now if ω is a p-form,

$$\begin{aligned} \omega &= \sum a_H(\mathbf{x})\, dx^H, \\ d\omega &= \sum d(a_H\, dx^H) \\ &= \sum (da_H)\, dx^H \\ &= \sum \frac{\partial a_H}{\partial x^j}\, dx^j\, dx^H, \end{aligned}$$

which shows that the recipe (i–iv) completely determines $d\omega$. To prove that there exists such an operator d, we simply set

$$d\omega = \sum \frac{\partial a_H}{\partial x^j}\, dx^j\, dx^H$$

for $\omega = \sum a_H\, dx^H$ and check that the properties are satisfied. Properties (i) and (iv) are fairly clear; let us look at (ii) and (iii). Evidently if we can establish these for monomials, by summation they will follow generally.

Suppose

$$\lambda = a\, dx^H, \qquad \mu = b\, dx^K.$$

Then

$$\begin{aligned} d(\lambda \wedge \mu) &= d(ab\, dx^H\, dx^K) \\ &= \sum \frac{\partial(ab)}{\partial x^i}\, dx^i\, dx^H\, dx^K \\ &= \sum \frac{\partial a}{\partial x^i}\, b\, dx^i\, dx^H\, dx^K + \sum a \frac{\partial b}{\partial x^i}\, dx^i\, dx^H\, dx^K \\ &= \sum \left(\frac{\partial a}{\partial x^i}\, dx^i\, dx^H\right) \wedge (b\, dx^K) \\ &\quad + (-1)^{(\deg \lambda)} \sum (a\, dx^H) \wedge \left(\frac{\partial b}{\partial x^i}\, dx^i\, dx^K\right) \\ &= (d\lambda) \wedge \mu + (-1)^{(\deg \lambda)} \lambda \wedge d\mu. \end{aligned}$$

The sign results from

$$dx^i\,dx^H = (-1)^{(\deg \lambda)}\,dx^H\,dx^i.$$

This proves (ii).

Again, let $\omega = a\,dx^H$. Then

$$\begin{aligned} d(d\omega) &= d\left(\sum \frac{\partial a}{\partial x^i}\,dx^i\,dx^H\right) \\ &= \sum \frac{\partial^2 a}{\partial x^i\,\partial x^j}\,dx^j\,dx^i\,dx^H \\ &= \frac{1}{2}\sum\left(\frac{\partial^2 a}{\partial x^i\,\partial x^j} - \frac{\partial^2 a}{\partial x^j\,\partial x^i}\right)dx^j\,dx^i\,dx^H \\ &= 0, \end{aligned}$$

which verifies (iii).

Property (iii) is nothing more than the equality of mixed second partial derivatives. It is the source of most "integrability conditions" in partial differential equations and differential geometry. It is usually referred to as the *Poincaré Lemma.*

3.3. Mappings

We study the following situation: $\mathbf{U}$ is a domain in $\mathbf{E}^m$, $\mathbf{V}$ is a domain in $\mathbf{E}^n$ and ϕ is a smooth mapping on $\mathbf{U}$ into $\mathbf{V}$. We write

$$\phi\colon\quad \mathbf{U} \longrightarrow \mathbf{V}.$$

Also, we denote by $x^1, \cdots, x^m$ the coordinates of $\mathbf{E}^m$ and by $y^1, \cdots, y^n$ the coordinates of $\mathbf{E}^n$. Then we can write

$$y^i = y^i(x^1, \cdots, x^m)$$

to show that the point with coordinates $\mathbf{x}$ is transformed by ϕ to the point with coordinates $\mathbf{y}$. The functions $y^i(\mathbf{x})$ are smooth.

As before, $\mathbf{R}$ denotes the reals. If g is any real-valued function on $\mathbf{V}$,

$$g\colon\quad \mathbf{V} \longrightarrow \mathbf{R}$$

then we may combine this with ϕ to obtain a function on $\mathbf{U}$ to $\mathbf{R}$ which we write

$$\phi^* g = g \circ \phi.$$

Thus

$$\phi^*\colon\quad \mathbf{F}^0(\mathbf{V}) \longrightarrow \mathbf{F}^0(\mathbf{U}).$$

From the mapping ϕ on $\mathbf{U}$ to $\mathbf{V}$ we have constructed a new (induced) mapping ϕ^* on $\mathbf{F}^0(\mathbf{V})$ to $\mathbf{F}^0(\mathbf{U})$.

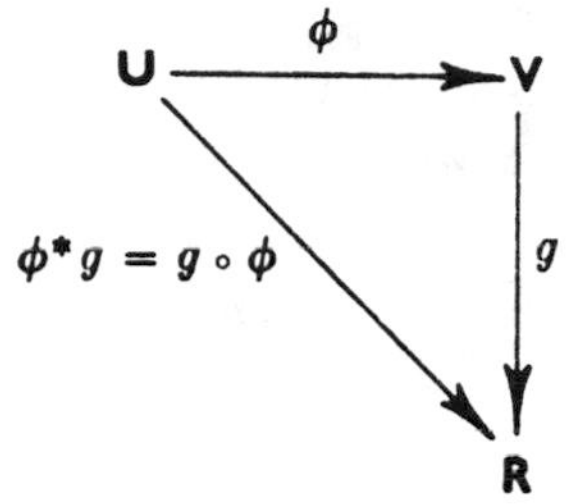

We are now going to define a map ϕ^* taking p-forms on $\mathbf{V}$ to p-forms on $\mathbf{U}$:

$$\phi^*: \quad \mathbf{F}^p(\mathbf{V}) \longrightarrow \mathbf{F}^p(\mathbf{U}).$$

(Strictly speaking we should index ϕ^* and write $\phi_p{}^*$, $p = 0, 1, \cdots$, but we shall skip this.) We have taken care of $p = 0$ already. The crucial case is $p = 1$; after we do that, the algebraic considerations of Chapter II do the rest of the work.

The basic idea is *substitution of coordinate functions*, replacing dy^i by

$$\sum \frac{\partial y^i}{\partial x^j}\, dx^j.$$

Thus if $\omega = \sum a_i(\mathbf{y})\, dy^i$ is a one-form on $\mathbf{V}$, we set

$$\phi^* \omega = \sum a_i(\mathbf{y}(\mathbf{x}))\, \frac{\partial y^i}{\partial x^j}\, dx^j.$$

We now have

$$\phi^*: \quad \mathbf{F}^1(\mathbf{V}) \longrightarrow \mathbf{F}^1(\mathbf{U}).$$

By the method of Section 2.4, we extend this mapping to the exterior products to obtain

$$\phi^*: \quad \mathbf{F}^p(\mathbf{V}) \longrightarrow \mathbf{F}^p(\mathbf{U}).$$

As an example,

$$\begin{aligned}
\phi^* (dy^1\, dy^2) &= (\phi^* dy^1)(\phi^* dy^2) \\
&= \left(\sum \frac{\partial y^1}{\partial x^i}\, dx^i\right)\left(\sum \frac{\partial y^2}{\partial x^j}\, dx^j\right) \\
&= \sum \frac{\partial y^1}{\partial x^i}\, \frac{\partial y^2}{\partial x^j}\, dx^i\, dx^j \\
&= \frac{1}{2} \sum \left(\frac{\partial y^1}{\partial x^i}\, \frac{\partial y^2}{\partial x^j} - \frac{\partial y^1}{\partial x^j}\, \frac{\partial y^2}{\partial x^i}\right) dx^i\, dx^j \\
&= \frac{1}{2} \sum \frac{\partial(y^1, y^2)}{\partial(x^i, x^j)}\, dx^i\, dx^j.
\end{aligned}$$

We now list the basic properties of ϕ^*.

(i) $\phi^*(\omega + \eta) = \phi^*\omega + \phi^*\eta$
(ii) $\phi^*(\lambda \wedge \mu) = (\phi^*\lambda) \wedge (\phi^*\mu)$
(iii) If ω is a p-form on $\mathbf{V}$,

$$d(\phi^*\omega) = \phi^*(d\omega)$$

(iv) If $\phi\colon\ \mathbf{U} \longrightarrow \mathbf{V}$ and $\psi\colon\ \mathbf{V} \longrightarrow \mathbf{W}$, then

$$(\psi \circ \phi)^* = \phi^* \circ \psi^*.$$

The first property is evident and the second follows from the final formula of Section 2.4.

Property (iii) is essentially the chain rule for partial derivatives. First we take a 0-form g on $\mathbf{V}$.

$$dg = \sum \frac{\partial g}{\partial y^j}\, dy^j,$$

$$\phi^* dg = \sum \frac{\partial g(\mathbf{y}(\mathbf{x}))}{\partial y^j} \frac{\partial y^j}{\partial x^i}\, dx^i$$

$$= \sum \frac{\partial(\phi^* g)}{\partial x^i}\, dx^i = d\,\phi^* g.$$

We proceed inductively, supposing we have verified (iii) for $(p-1)$-forms. It suffices to verify (iii) for p-forms ω which are monomial since each p-form is a sum of such. Suppose then that

$$\omega = g\, dy^H = g\, d\eta$$

where $\eta = y^{h_1}\, dy^{h_2} \cdots dy^{h_p}$ is a $(p-1)$-form. Then

$$\phi^*\omega = (\phi^* g)(\phi^* d\eta) = (\phi^* g) \wedge (d\,\phi^*\eta),$$

$$d(\phi^*\omega) = d(\phi^* g) \wedge d(\phi^*\eta),$$

and

$$d\omega = dg \wedge d\eta,$$

$$\phi^* d\omega = (\phi^* dg) \wedge (\phi^* d\eta)$$

$$= d(\phi^* g) \wedge d(\phi^*\eta) = d\,\phi^*\omega;$$

we have pushed through the next case.

We now look at the final property (iv).

For a 0-form (function) h on $\mathbf{W}$ we have

$$[(\psi \circ \phi)^* h](\mathbf{x}) = h[(\psi \circ \phi)(\mathbf{x})] = h\{\psi[\phi(\mathbf{x})]\}$$

$$= [\psi^* h][\phi(\mathbf{x})] = \{\phi^*[\psi^* h]\}(\mathbf{x})$$

$$= [(\phi^* \circ \psi^*)h](\mathbf{x}),$$

hence

$$(\psi \circ \phi)^* h = (\phi^* \circ \psi^*) h.$$

An induction similar to that above establishes the property in general. All it means is that one can substitute directly the expressions for the coordinates

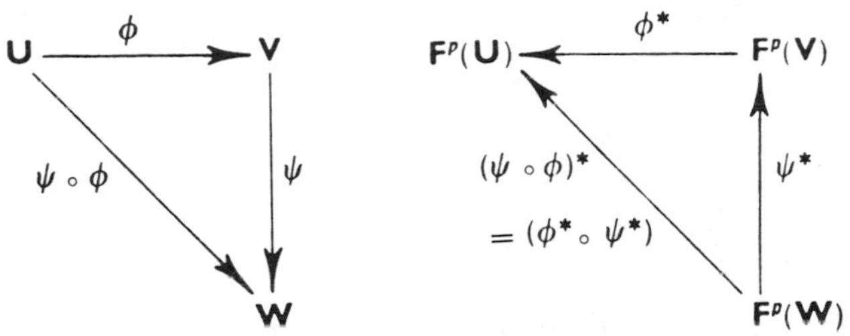

z^k on **W** in terms of the coordinates x^i on **U**, or indirectly by first going through the coordinates y^j of **V**; the results are the same.

What has really been seen in this section is that one can carry on fearlessly with the most obvious kind of calculations with differential forms.

Examples. Consider the map ϕ: $t \longrightarrow (x, y)$ on $\mathbf{E}^1 \longrightarrow \mathbf{E}^2$ given by $x = t^2, y = t^3$. If $\omega = x\,dy$, a one-form on $\mathbf{E}^2$,

$$\phi^* \omega = (t^2) \frac{\partial y}{\partial t}\, dt = 3t^4\, dt.$$

Take the map ψ: $(x, y) \longrightarrow t = x - y$.

$$\psi^* (dt) = dx - dy.$$

One final remark. Suppose $m < n$ and ϕ is a map on the domain **U** of $\mathbf{E}^m$ into the domain **V** of $\mathbf{E}^n$. If ω is a p-form on **V** and $p > m$, then necessarily $\phi^* \omega = 0$.

3.4. Change of Coordinates

We apply the results of the last section to the special case in which **U** and **V** are both domains in $\mathbf{E}^n$ and ϕ is a one-to-one mapping on **U** onto **V** with both ϕ and $\psi = \phi^{-1}$ smooth. (Note the map $x \longrightarrow y = x^3$ on $\mathbf{E}^1 \longrightarrow \mathbf{E}^1$ is one-to-one and smooth. But the inverse map $y \longrightarrow x = y^{1/3}$ is not smooth—no derivative at $y = 0$.) In each figure, ι is the identity map, $\iota(\mathbf{x}) = \mathbf{x}$. It follows that ϕ^* is a one-one map on $\mathbf{F}^p(\mathbf{V})$ onto $\mathbf{F}^p(\mathbf{U})$ and its inverse is ψ^*. If we interpret the coordinates $\mathbf{y}$ of **V** as new coordinates on **U**, the result

$$d\phi^* \omega = \phi^* d\omega$$

means that *the exterior derivative of a differential form is independent of the coordinate system in which it is computed.*

This inner consistency of the differential form calculus is most important. Later we shall base the global theory (forms on manifolds) on this.

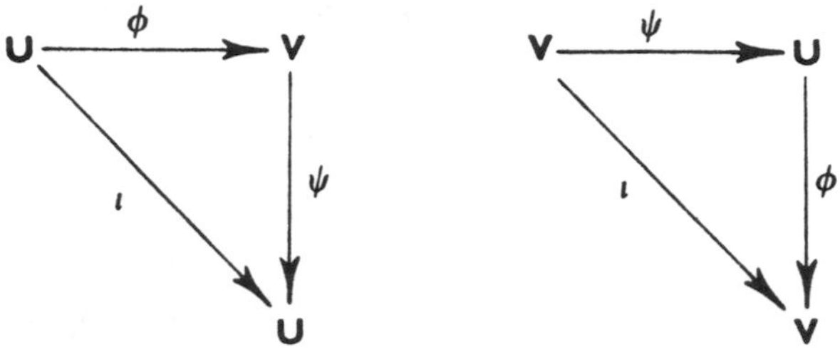

We note in passing that with a proper formulation this independence of d on the coordinate system can be obtained as a consequence of the four basic defining properties (i–iv) of the exterior derivative in Section 3.2.

3.5. An Example from Mechanics

The following problem is taken from É. Goursat [15, p. 85]. We work in a region with coordinates $(\mathbf{x}, \mathbf{u}) = (x_1, \cdots, x_n, u_1, \cdots, u_n)$. We are given a function

$$\phi = \phi(\mathbf{x}, \mathbf{u})$$

which is supposed homogeneous of degree 2 in the variable $\mathbf{u}$. (For example, a kinetic energy form $\sum a_{ij}(\mathbf{x})\, u_i u_j$.) Define

$$p_i = \partial\phi/\partial u_i .$$

We assume that the mapping $(\mathbf{x}, \mathbf{u}) \longrightarrow (\mathbf{x}, \mathbf{p})$ defines a regular change of variables. We then write

$$\phi(\mathbf{x}, \mathbf{u}) = \psi(\mathbf{x}, \mathbf{p}).$$

The problem is to prove the relations

$$\frac{\partial\psi}{\partial x_i} = -\frac{\partial\phi}{\partial x_i}, \qquad \frac{\partial\psi}{\partial p_k} = u_k .$$

The proof depends on two things, the Euler formula for homogeneous functions which in our case implies

$$\sum \frac{\partial\phi}{\partial u_k} u_k = 2\phi,$$

i.e.,

$$\sum p_k u_k = 2\phi,$$

and the fact that exterior relations are independent of how they are derived.

We have

$$d\phi = \sum \frac{\partial \phi}{\partial x_i} dx_i + \sum \frac{\partial \phi}{\partial u_k} du_k = \sum \frac{\partial \phi}{\partial x_i} dx_i + \sum p_k du_k$$

and

$$2d\phi = \sum p_k du_k + \sum u_k dp_k,$$

hence by subtracting

$$d\phi = -\sum \frac{\partial \phi}{\partial x_i} dx_i + \sum u_k dp_k.$$

Now everything follows from $\phi = \psi$ and

$$d\psi = \sum \frac{\partial \psi}{\partial x_i} dx_i + \sum \frac{\partial \psi}{\partial p_k} dp_k.$$

3.6. Converse of the Poincaré Lemma

The Poincaré Lemma, $d(d\omega) = 0$, has these interpretations in 3-space:

$$\text{curl}\,(\text{grad}\, f) = 0$$

$$\text{div}\,(\text{curl}\, \mathbf{v}) = 0$$

according to the examples at the beginning of Section 3.2. In vector analysis one proves that a curl-free vector field is a gradient by line integrals and that a divergence-free vector field is a curl, usually by a brute force method. We are now going to prove a general result. *If ω is a p-form ($p \geqq 1$) and $d\omega = 0$, then there is a $(p-1)$-form α such that $\omega = d\alpha$.* The result is hard if $p > 1$ because there are many solutions. Also the result is valid only in domains which are not too complicated topologically.

The demonstration is based on a "cylinder construction." We begin with a domain $\mathbf{U}$ in $\mathbf{E}^n$. We denote by $\mathbf{I} = [0, 1]$ the unit interval on the t-axis and consider the *cylinder* or product space.

$$\mathbf{I} \times \mathbf{U}.$$

This consists of all pairs $(t, \mathbf{x})$ where $0 \leqq t \leqq 1$ and $\mathbf{x}$ runs over points of $\mathbf{U}$.

We single out the two maps which identify $\mathbf{U}$ with the top and bottom of the cylinder, namely,

$$j_1\colon \quad \mathbf{U} \longrightarrow \mathbf{I} \times \mathbf{U}, \qquad j_1(\mathbf{x}) = (1, \mathbf{x})$$

$$j_0\colon \quad \mathbf{U} \longrightarrow \mathbf{I} \times \mathbf{U}, \qquad j_0(\mathbf{x}) = (0, \mathbf{x}).$$

Thus

$$j_i^*\colon \quad \mathbf{F}^p(\mathbf{I} \times \mathbf{U}) \longrightarrow \mathbf{F}^p(\mathbf{U}) \qquad (i = 0, 1).$$

For example, to form $j_1^*\omega$ where ω is a form on $\mathbf{I} \times \mathbf{U}$, simply replace t by 1 wherever it occurs in ω (and dt by 0 correspondingly).

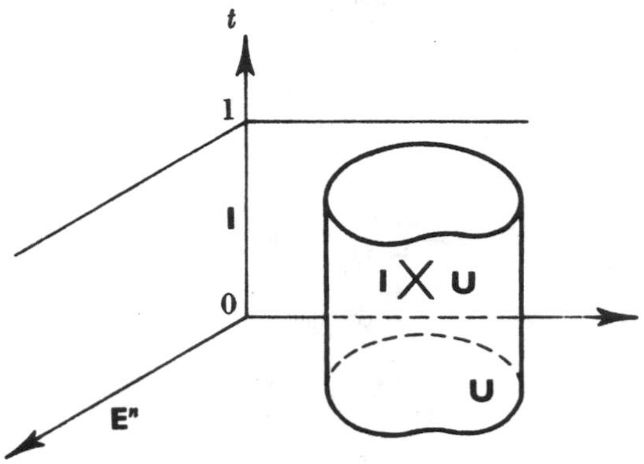

We now form a new operation K,

$$K: \quad \mathbf{F}^{p+1}(\mathbf{I} \times \mathbf{U}) \longrightarrow \mathbf{F}^{p}(\mathbf{U});$$

K is defined on monomials by the formulas

$$K(a(t, \mathbf{x})\, dx^H) = 0$$

$$K(a(t, \mathbf{x})\, dt\, dx^J) = \left(\int_0^1 a(t, \mathbf{x})\, dt\right) dx^J,$$

and on general differential forms by summing the results on the monomial parts. Here is the basic property of K: *If* ω *is any* $(p+1)$*-form on* $\mathbf{I} \times \mathbf{U}$, *then*

$$K(d\omega) + d(K\omega) = j_1{}^*\omega - j_0{}^*\omega.$$

It is enough to check this for monomials.

Case 1. $\omega = a(t, \mathbf{x})\, dx^H$.

We have $K\omega = 0$, $dK\omega = 0$,

$$d\omega = \frac{\partial a}{\partial t}\, dt\, dx^H + [\text{terms free of } dt],$$

$$K\, d\omega = \left(\int_0^1 \frac{\partial a}{\partial t}\, dt\right) dx^H = [a(1, \mathbf{x}) - a(0, \mathbf{x})]\, dx^H.$$

But $j_1{}^*\omega = a(1, \mathbf{x})\, dx^H$, $j_0{}^*\omega = a(0, \mathbf{x})\, dx^H$, so the formula is valid.

Case 2. $\omega = a(t, \mathbf{x})\, dt\, dx^J$.

First $j_1{}^*\omega = j_0{}^*\omega = 0$. Next,

$$K\,d\omega = K\left[-\sum \frac{\partial a}{\partial x^i}\,dt\,dx^i\,dx^J\right]$$

$$= -\sum \left(\int_0^1 \frac{\partial a}{\partial x^i}\,dt\right) dx^i\,dx^J,$$

$$dK\omega = d\left[\left(\int_0^1 a(t, \mathbf{x})\,dt\right) dx^J\right]$$

$$= \sum \frac{\partial}{\partial x^i}\left[\int_0^1 a(t, \mathbf{x})\,dt\right] dx^i\,dx^J$$

$$= \sum \left(\int_0^1 \frac{\partial a}{\partial x^i}\,dt\right) dx^i\,dx^J,$$

so the formula again works.

Definition. A domain **U** is *deformable to a point* P if there is a mapping

$$\phi\colon\ \mathbf{I} \times \mathbf{U} \longrightarrow \mathbf{U}$$

such that

$$\phi(1, \mathbf{x}) = \mathbf{x},$$

$$\phi(0, \mathbf{x}) = P.$$

The boundary conditions may be interpreted in terms of the j_i as follows:

$$\phi \circ j_1 = \iota, \qquad \phi \circ j_0 = P.$$

For a $(p + 1)$-form ω on **U** we have as a consequence

$$j_1{}^*[\phi^*\,\omega] = \omega, \qquad j_0{}^*[\phi^*\,\omega] = 0.$$

Now we can state and prove the main result.

Let **U** *be a domain in* $\mathbf{E}^n$ *which can be deformed to a point* P. *Let* ω *be a* $(p + 1)$*-form on* **U** *such that* $d\omega = 0$. *Then there is a* p*-form* α *on* **U** *such that*

$$\omega = d\alpha.$$

We merely substitute $\phi^*\,\omega$ in the formula above to have

$$K[d(\phi^*\,\omega)] + d[K(\phi^*\,\omega)] = \omega.$$

But $d(\phi^*\,\omega) = \phi^*\,(d\omega) = 0$, hence $\omega = d\alpha$ with $\alpha = K(\phi^*\,\omega)$.

It is interesting to see how far the solution of the equation $d\alpha = \omega$ is determined. If β is another solution, then $d\beta = \omega = d\alpha$, $d(\alpha - \beta) = 0$. If $p \geqq 1$, we conclude by the main result again that $\alpha - \beta = d\lambda$ where λ is a $(p - 1)$-form. In other words, given one solution α, the general solution is $\alpha - d\lambda$ where λ is absolutely arbitrary. (When $p = 0$, α and β are functions and we conclude that $\alpha - \beta$ is constant.)

3.7. An Example

We shall illustrate this whole method in the case $n = 3$, $p = 2$. Thus we take a two-form

$$\omega = A\,dy\,dz + B\,dz\,dx + C\,dx\,dy$$

in $\mathbf{E}^3$ for which $d\omega = 0$, i.e.,

$$\frac{\partial A}{\partial x} + \frac{\partial B}{\partial y} + \frac{\partial C}{\partial z} = 0.$$

The space $\mathbf{E}^3$ can be deformed to 0 by the map

$$\phi(t, x, y, z) = (tx, ty, tz).$$

The assertion is that $\omega = d\alpha$ where

$$\alpha = K\phi^*\omega.$$

First we compute $\phi^*\omega$:

$$\begin{aligned}\phi^*\omega &= A(tx, ty, tz)\,d(ty)\,d(tz) + \cdots \\ &= A(tx, ty, tz)(t\,dy + y\,dt)(t\,dz + z\,dt) + \cdots \\ &= A(tx, ty, tz)(yt\,dt\,dz - zt\,dt\,dy) + \cdots + (\text{terms free of } dt).\end{aligned}$$

Now we have

$$\begin{aligned}\alpha = K(\phi^*\omega) = &\left(\int_0^1 A(tx, ty, tz)\,t\,dt\right)(y\,dz - z\,dy) \\ &+ \left(\int_0^1 B(tx, ty, tz)\,t\,dt\right)(z\,dx - x\,dz) \\ &+ \left(\int_0^1 C(tx, ty, tz)\,t\,dt\right)(x\,dy - y\,dx).\end{aligned}$$

One verifies after some calculation that indeed $d\alpha = \omega$.

3.8. Further Remarks

For

$$\omega = A\,dy\,dz + B\,dz\,dx + C\,dx\,dy$$

the problem of finding

$$\alpha = P\,dx + Q\,dy + R\,dz$$

so that

$$d\alpha = \omega$$

is that of finding three unknown functions P, Q, R of the three variables x, y, z so that the system

$$\begin{cases} \dfrac{\partial R}{\partial y} - \dfrac{\partial Q}{\partial z} = A \\ \dfrac{\partial P}{\partial z} - \dfrac{\partial R}{\partial x} = B \\ \dfrac{\partial Q}{\partial x} - \dfrac{\partial P}{\partial y} = C \end{cases}$$

of three partial differential equations is satisfied, the given functions A, B, C being subject to the necessary condition

$$\frac{\partial A}{\partial x} + \frac{\partial B}{\partial y} + \frac{\partial C}{\partial z} = 0.$$

It is remarkable that this system (and the more general ones covered in Section 3.6) can be solved by an explicit formula involving quadratures. In general, the theory of exterior differential forms exposes many types of systems of partial differential equations which are reducible to systems of ordinary differential equations and often solved by quadratures.

Another point to be noted is this. If we are dealing with a $(p+1)$-form ω such that $d\omega = 0$ and ω happens to depend on several parameters smoothly, then we can find an α such that $d\alpha = \omega$ and α depends on the same parameters just as smoothly. This again follows from the explicit formulas of Section 3.6.

3.9. Problems

1. Let

$$\omega = \tfrac{1}{2} \sum a_{ij}\, dx^i\, dx^j, \qquad a_{ij} + a_{ji} = 0.$$

Prove that

$$d\omega = \tfrac{1}{6} \sum \left(\frac{\partial a_{ij}}{\partial x^k} + \frac{\partial a_{jk}}{\partial x^i} + \frac{\partial a_{ki}}{\partial x^i} \right) dx^i\, dx^j\, dx^k.$$

2. Consider a linear transformation ϕ: $\mathbf{E}^n \longrightarrow \mathbf{E}^n$, $\phi(x^1, \cdots, x^n) = (y^1, \cdots, y^n)$, where

$$y^i = \sum a^i{}_j x^j + b^i, \qquad a^i{}_j,\ b^i \text{ constant}.$$

What is $\phi^*(dx^1 \cdots dx^n)$?

3. Consider the mapping

$$\phi\colon \quad (x, y) \longrightarrow (xy, 1)$$

on $\mathbf{E}^2$ into $\mathbf{E}^2$. Compute $\phi^*(dx)$, $\phi^*(dy)$, and $\phi^*(y\,dx)$.

4. Complete the unfinished calculation of Section 3.7.

IV

Applications

4.1. Moving Frames in E^3

We first point out that in dealing with vectors in Euclidean space, no matter where we draw them for picturesque purposes, when we deal with them analytically, they always start at the origin.

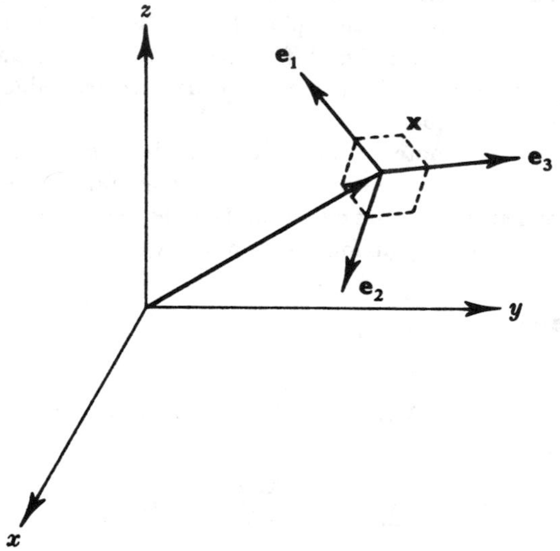

We attach to each point $\mathbf{x}$ of $\mathbf{E}^3$ a right-handed orthonormal frame $\mathbf{e}_1, \mathbf{e}_2, \mathbf{e}_3$ and suppose that the vector fields $\mathbf{e}_i$ are smooth fields.

What we shall do is express everything in sight in terms of the $\mathbf{e}_i$, apply d to these relations to derive further ones, and continue until we obtain no further results.

First of all, $d\mathbf{x}$ is a vector with one-form coefficients, for example, $d\mathbf{x} = (dx, dy, dz) = dx\,\mathbf{i} + dy\,\mathbf{j} + dz\,\mathbf{k}$. We express $d\mathbf{x}$ in terms of the frame $\mathbf{e}_1, \mathbf{e}_2, \mathbf{e}_3$ at the point $\mathbf{x}$, which we certainly may do, say, by first expanding $\mathbf{i}, \mathbf{j}, \mathbf{k}$ in terms of the $\mathbf{e}_i$ and then collecting terms:

$$d\mathbf{x} = \sigma_1 \mathbf{e}_1 + \sigma_2 \mathbf{e}_2 + \sigma_3 \mathbf{e}_3,$$

where the σ_i are one-forms. We do the same with each $\mathbf{e}_i$:

$$d\mathbf{e}_i = \omega_{i1}\mathbf{e}_1 + \omega_{i2}\mathbf{e}_2 + \omega_{i3}\mathbf{e}_3 \qquad (i = 1, 2, 3)$$

where the ω_{ij} are one-forms.

Since $\mathbf{e}_i \cdot \mathbf{e}_k = \delta_{ik}$, we have

$$d\mathbf{e}_i \cdot \mathbf{e}_k + \mathbf{e}_i \cdot d\mathbf{e}_k = 0,$$

that is,

$$\omega_{ik} + \omega_{ki} = 0.$$

In particular, $\omega_{ii} = 0$.

It will be convenient to introduce some matrix notation. We set

$$\mathbf{e} = \begin{pmatrix} \mathbf{e}_1 \\ \mathbf{e}_2 \\ \mathbf{e}_3 \end{pmatrix}, \qquad \boldsymbol{\sigma} = (\sigma_1, \sigma_2, \sigma_3), \qquad \Omega = \|\omega_{ij}\|$$

and have these structure equations:

$$d\mathbf{x} = \boldsymbol{\sigma}\mathbf{e},$$
$$d\mathbf{e} = \Omega\mathbf{e},$$
$$\Omega + {}^t\Omega = 0.$$

Here applying d to a matrix means simply applying it to each element. In the last equation, the left-hand superscript t denotes *transpose* of the matrix, i.e., interchange of rows and columns, so this equation expresses the skew-symmetry of Ω.

From $d(d\mathbf{x}) = 0$ we have

$$d\boldsymbol{\sigma}\,\mathbf{e} - \boldsymbol{\sigma}\,d\mathbf{e} = 0,$$
$$d\boldsymbol{\sigma}\,\mathbf{e} - \boldsymbol{\sigma}\Omega\,\mathbf{e} = 0,$$
$$(d\boldsymbol{\sigma} - \boldsymbol{\sigma}\Omega)\,\mathbf{e} = 0.$$

Because the $\mathbf{e}_i$ are linearly independent, this means

$$d\boldsymbol{\sigma} = \boldsymbol{\sigma}\Omega.$$

Similarly, from $d(d\mathbf{e}) = 0$, we have

$$0 = d\Omega\,\mathbf{e} - \Omega\,d\mathbf{e} = (d\Omega - \Omega^2)\,\mathbf{e},$$
$$d\Omega = \Omega^2.$$

In summary, then we have

Structure equations	Integrability conditions
$\left\{\begin{aligned} d\mathbf{x} &= \boldsymbol{\sigma}\mathbf{x} \\ d\mathbf{e} &= \Omega\mathbf{e} \\ \Omega + {}^t\Omega &= 0 \end{aligned}\right\}$	$\left\{\begin{aligned} d\boldsymbol{\sigma} &= \boldsymbol{\sigma}\Omega \\ d\Omega &= \Omega^2 \end{aligned}\right\}$

Further differentiation does not lead to new results. We shall see in our study of Riemannian geometry that the equation $d\Omega - \Omega^2 = 0$ expresses the lack of curvature of Euclidean space.

A point to be noticed is that the three-form $\sigma_1 \wedge \sigma_2 \wedge \sigma_3$ is precisely the element of volume in $\mathbf{E}^3$:

$$\sigma_1 \wedge \sigma_2 \wedge \sigma_3 = dx\,dy\,dz.$$

We shall verify this in the next section.

It will be observed that the calculations of this section work equally well in $\mathbf{E}^n$.

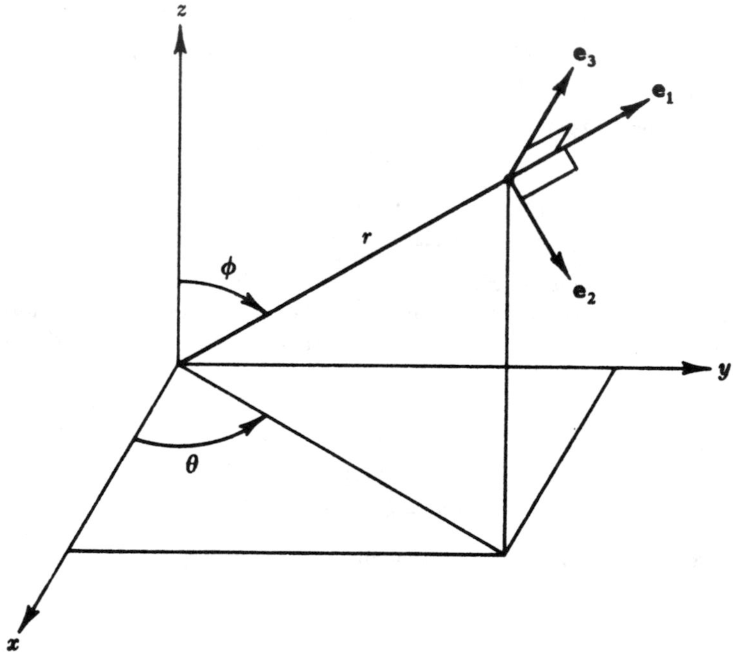

Example. Spherical coordinates. The orthonormal unit vectors $\mathbf{e}_1$, $\mathbf{e}_2$, $\mathbf{e}_3$ are taken in the directions of increasing r, ϕ, θ, respectively. From

$$\mathbf{x} = (r \sin\phi \cos\theta,\ r\sin\phi \sin\theta,\ r\cos\phi)$$

we have

$$\begin{aligned} d\mathbf{x} &= (\sin\phi\cos\theta,\ \sin\phi\sin\theta,\ \cos\phi)\,dr \\ &\quad + (r\cos\phi\cos\theta,\ r\cos\phi\sin\theta,\ -r\sin\phi)\,d\phi \\ &\quad + (-r\sin\phi\sin\theta,\ r\sin\phi\cos\theta,\ 0)\,d\theta \\ &= (dr)\,\mathbf{e}_1 + (r\,d\phi)\,\mathbf{e}_2 + (r\sin\phi\ d\theta)\,\mathbf{e}_3 \end{aligned}$$

with

$$\begin{cases} \mathbf{e}_1 = (\sin\phi\cos\theta,\ \sin\phi\sin\theta,\ \cos\phi) \\ \mathbf{e}_2 = (\cos\phi\cos\theta,\ \cos\phi\sin\theta,\ -\sin\phi) \\ \mathbf{e}_3 = (-\sin\theta,\ \cos\theta,\ 0) \end{cases}$$

and so

$$\sigma_1 = dr, \qquad \sigma_2 = r\,d\phi, \qquad \sigma_3 = r\sin\phi\,d\theta.$$

Differentiating,

$$d\mathbf{e}_1 = (d\phi)\,\mathbf{e}_2 + (\sin\phi\,d\theta)\,\mathbf{e}_3$$

$$d\mathbf{e}_2 = (-d\phi)\,\mathbf{e}_1 + (\cos\phi\,d\theta)\,\mathbf{e}_3$$

hence since Ω is skew-symmetric,

$$\Omega = \begin{pmatrix} 0 & d\phi & \sin\phi\,d\theta \\ -d\phi & 0 & \cos\phi\,d\theta \\ -\sin\phi\,d\theta & -\cos\phi\,d\theta & 0 \end{pmatrix}.$$

The volume element is

$$\sigma_1 \wedge \sigma_2 \wedge \sigma_3 = r^2 \sin\phi\,dr\,d\phi\,d\theta.$$

4.2. Relation between Orthogonal and Skew-symmetric Matrices

It is no accident that Ω turns out to be skew-symmetric. This is a consequence of the principle that the first-order approximation to an orthogonal transformation is a skew-symmetric one. We shall look at this from several viewpoints.

A matrix B is orthogonal if its transpose equals its inverse, ${}^tB = B^{-1}$, or $B\,{}^tB = {}^tBB = I$. Suppose A is skew-symmetric, $A + {}^tA = 0$. Then for small ε we set $B = I + \varepsilon A$ and have

$$B\,{}^tB = (I + \varepsilon A)(I - \varepsilon A) = I + O(\varepsilon^2)$$

so that B is orthogonal up to first-order terms.

Here is another approach. Let A be skew-symmetric. Since the characteristic roots of A are pure imaginary, $I + A$ and $I - A$ are nonsingular. Set

$$B = \frac{I + A}{I - A}.$$

Then

$$B\,{}^tB = \left(\frac{I + A}{I - A}\right)\left(\frac{I - A}{I + A}\right) = I$$

so that B is orthogonal.

Next we re-examine the calculations of the last section. Let

$$\mathbf{i} = \begin{pmatrix} \mathbf{i}_1 \\ \mathbf{i}_2 \\ \mathbf{i}_3 \end{pmatrix}$$

where the $\mathbf{i}_j$ are the fixed unit vectors in the x, y, z directions, respectively ($\mathbf{i}$, $\mathbf{j}$, $\mathbf{k}$ in usual vector notation). Then

$$\mathbf{e}_i = \sum b_{ij}\mathbf{i}_j, \qquad \mathbf{e} = B\mathbf{i}$$

leading to a matrix $B = \|b_{ij}\|$ which is clearly orthogonal:

$$I = \mathbf{e}\,{}^t\mathbf{e} = B\mathbf{i}\,{}^t\mathbf{i}\,{}^tB = BI\,{}^tB = B\,{}^tB.$$

(Now we can prove the fact $dx\,dy\,dz = \sigma_1 \wedge \sigma_2 \wedge \sigma_3$ mentioned at the end of the last section. We have

$$d\mathbf{x} = (dx, dy, dz)\,\mathbf{i} = \boldsymbol{\sigma}\mathbf{e} = \boldsymbol{\sigma}B\mathbf{i},$$
$$(dx, dy, dz) = \boldsymbol{\sigma}B,$$

hence

$$dx\,dy\,dz = |B|\sigma_1 \wedge \sigma_2 \wedge \sigma_3\,.$$

But from ${}^tB\,B = I$ we have $|B|^2 = 1$, $|B| = \pm 1$. Since we are supposing $\mathbf{e}$ is a right-handed system, $|B| = +1$,

$$dx\,dy\,dz = \sigma_1 \wedge \sigma_2 \wedge \sigma_3\,.)$$

Then we have

$$d\mathbf{e} = dB\,\mathbf{i} = (dB)\,B^{-1}\mathbf{e}$$

so that

$$\Omega = (dB)\,B^{-1}.$$

We note this general result: *If A is an orthogonal matrix whose elements are functions of any number of variables, then*

$$(dA)\,A^{-1}$$

is a skew-symmetric matrix of one-forms.

For we have

$$\begin{aligned} {}^tA\,A &= I, \\ {}^tdA\,A + {}^tA\,dA &= 0, \\ {}^tA^{-1}\,{}^tdA + dA\,A^{-1} &= 0, \\ {}^t(dA\,A^{-1}) + dA\,A^{-1} &= 0. \end{aligned}$$

There is also a converse which is important. *Suppose A is a matrix of functions defined on a domain $\mathbf{U}$. Suppose A is orthogonal at a single point of $\mathbf{U}$ and that*

$$dA = \Lambda A$$

where Λ *is a skew-symmetric matrix of one-forms. Then* A *is orthogonal on all of* $\mathbf{U}$.

We set $C = {}^tA\,A$ and have

$$dC = ({}^t dA)A + {}^tA(dA) = (-{}^tA\,\Lambda)A + {}^tA(\Lambda A) = 0,$$

hence C is a constant matrix on $\mathbf{U}$. But we are assuming $C = I$ at one point of $\mathbf{U}$, hence $C = I$ on $\mathbf{U}$, ${}^tA\,A = I$ on $\mathbf{U}$, A is orthogonal.

Another point is this. If A is a variable orthogonal matrix (transformation), each point $\mathbf{v}_0$ of space is sent by the general A to

$$\mathbf{v} = A\mathbf{v}_0 .$$

We then have

$$d\mathbf{v} = dA\,\mathbf{v}_0 = (dA)A^{-1}\mathbf{v}$$

so that one passes from $\mathbf{v}$ to the "infinitely near" vector $\mathbf{v} + d\mathbf{v}$ under the action of the general A of our family by means of

$$\mathbf{v} \longrightarrow \mathbf{v} + d\mathbf{v} = [I + (dA)A^{-1}]\mathbf{v}$$

with the skew-symmetric $(dA)A^{-1}$ representing this "infinitesimal transformation."

All of these considerations work equally well in $\mathbf{E}^n$.

4.3. The 6-dimensional Frame Space

We consider the space of all right-handed orthonormal frames $\mathbf{E}_1, \mathbf{E}_2, \mathbf{E}_3$ at all points $\mathbf{x}$ of $\mathbf{E}^3$. This space is 6-dimensional because we have three degrees of freedom in choosing $\mathbf{x}$, two degrees of freedom in choosing the unit vector $\mathbf{E}_1$, one degree of freedom in choosing the unit vector $\mathbf{E}_2$ perpendicular to $\mathbf{E}_1$ and then $\mathbf{E}_3$ is determined.

We write

$$\mathbf{E} = \begin{pmatrix} \mathbf{E}_1 \\ \mathbf{E}_2 \\ \mathbf{E}_3 \end{pmatrix}$$

and have

$$\mathbf{E} = A\mathbf{e}$$

where A is a variable (three parameter) orthogonal matrix and $\mathbf{e} = \mathbf{e}(\mathbf{x})$ is a definite moving frame.

Then

$$d\mathbf{x} = \sigma\mathbf{e} = \sigma A^{-1}\mathbf{E},$$

$$d\mathbf{E} = (dA)\mathbf{e} + A\,d\mathbf{e} = [dA + A\Omega]\mathbf{e}$$
$$= [dA + A\Omega]A^{-1}\mathbf{E}.$$

We set

$$\tilde{\sigma} = \sigma A^{-1}, \qquad \tilde{\Omega} = (dA)A^{-1} + A\Omega A^{-1}.$$

These are matrices of one-forms on the 6-dimensional frame space and we have

Structure equations Integrability conditions

$$\begin{Bmatrix} d\mathbf{x} = \tilde{\sigma}\mathbf{E} \\ d\mathbf{E} = \tilde{\Omega}\mathbf{E} \\ \tilde{\Omega} + {}^t\tilde{\Omega} = 0 \end{Bmatrix} \qquad \begin{Bmatrix} d\tilde{\sigma} = \tilde{\sigma}\tilde{\Omega} \\ d\tilde{\Omega} = \tilde{\Omega}^2. \end{Bmatrix}$$

To check the integrability conditions we note

$$0 = d(d\mathbf{x}) = d\tilde{\sigma}\mathbf{E} - \tilde{\sigma}d\mathbf{E} = (d\tilde{\sigma} - \tilde{\sigma}\tilde{\Omega})\mathbf{E},\ d\tilde{\sigma} = \tilde{\sigma}\tilde{\Omega}, \quad \text{etc.}$$

In making a penetrating study of the differential geometry of $\mathbf{E}^3$ one is necessarily led to this 6-dimensional frame space and its differential forms $\tilde{\sigma}^i$, $\tilde{\omega}_{ij}$ which, it will be noted, are entirely independent of the choice of the moving frame $\mathbf{e}$ on $\mathbf{E}^3$.

4.4. The Laplacian, Orthogonal Coordinates

We continue the considerations of Sections 4.1 and 4.2. The forms dx, dy, dz make up an orthonormal basis for the Euclidean geometry of the space of one-forms at each point; these are related to the fixed (absolute) frame $\mathbf{i}$. From

$$\mathbf{e} = B\mathbf{i}, \qquad d\mathbf{x} = \sigma\mathbf{e} = (dx, dy, dz)\,\mathbf{i}$$

we have

$$\sigma B = (dx, dy, dz)$$

as already noted. As B is orthogonal, we see that $\sigma_1, \sigma_2, \sigma_3$ *is an orthonormal basis for one-forms at each point.*

Let f be a function on $\mathbf{E}^3$. Then we have

$$df = \frac{\partial f}{\partial x}dx + \frac{\partial f}{\partial y}dy + \frac{\partial f}{\partial z}dz,$$

$$*df = \frac{\partial f}{\partial x}dy\,dz + \frac{\partial f}{\partial y}dz\,dx + \frac{\partial f}{\partial z}dx\,dy,$$

$$d*df = \left(\frac{\partial^2 f}{\partial x^2} + \frac{\partial^2 f}{\partial y^2} + \frac{\partial^2 f}{\partial z^2}\right)dx\,dy\,dz = (\Delta f)\,dx\,dy\,dz.$$

The Laplacian Δf of f is known as soon as the three-form $d*df$ is known, for this has turned out to be the Laplacian multiplied by the volume element $dx\,dy\,dz$.

Now we know that the $*$ operator can be computed equally well in any orthonormal coordinate system. Also $\sigma_1 \wedge \sigma_2 \wedge \sigma_3 = dx\,dy\,dz$, so our procedure is this. We express df in terms of the σ_i,

$$df = a_1 \sigma_1 + a_2 \sigma_2 + a_3 \sigma_3 .$$

Then

$$*df = a_1 \sigma_2 \sigma_3 + a_2 \sigma_3 \sigma_1 + a_3 \sigma_1 \sigma_2 ,$$

$$d*df = (\Delta f)\sigma_1 \sigma_2 \sigma_3 .$$

A coordinate system u, v, w in a domain in $\mathbf{E}^3$ is called an *orthogonal coordinate* system if the vectors

$$\frac{\partial \mathbf{x}}{\partial u}, \qquad \frac{\partial \mathbf{x}}{\partial v}, \qquad \frac{\partial \mathbf{x}}{\partial w}$$

are mutually perpendicular. This means that for suitable functions λ, μ, ν, the vectors

$$\mathbf{e}_1 = \frac{1}{\lambda}\frac{\partial \mathbf{x}}{\partial u}, \qquad \mathbf{e}_2 = \frac{1}{\mu}\frac{\partial \mathbf{x}}{\partial v}, \qquad \mathbf{e}_3 = \frac{1}{\nu}\frac{\partial \mathbf{x}}{\partial w}$$

form an orthonormal, or moving frame. We shall presuppose that this is a right-handed one. (Otherwise we merely permute w and v.) We have

$$\begin{aligned} d\mathbf{x} &= du\,\frac{\partial \mathbf{x}}{\partial u} + dv\,\frac{\partial \mathbf{x}}{\partial v} + dw\,\frac{\partial \mathbf{x}}{\partial w} \\ &= (\lambda\,du)\,\mathbf{e}_1 + (\mu\,dv)\,\mathbf{e}_2 + (\nu\,dw)\,\mathbf{e}_3 \end{aligned}$$

so that

$$\sigma_1 = \lambda\,du, \qquad \sigma_2 = \mu\,dv, \qquad \sigma_3 = \nu\,dw$$

build an orthonormal frame for one-forms. Now we compute the Laplacian:

$$\begin{aligned} df &= f_u\,du + f_v\,dv + f_w\,dw \\ &= (f_u/\lambda)\sigma_1 + (f_v/\mu)\sigma_2 + (f_w/\nu)\sigma_3 . \\ *df &= (f_u/\lambda)\sigma_2\sigma_3 + (f_v/\mu)\sigma_3\sigma_1 + (f_w/\nu)\sigma_1\sigma_2 \\ &= (\mu\nu\,f_u/\lambda)\,dv\,dw + (\lambda\nu\,f_v/\mu)\,dw\,du + (\lambda\mu\,f_w/\nu)\,du\,dv. \end{aligned}$$

We compare this to

$$d*df = (\Delta f)\sigma_1\sigma_2\sigma_3 = \lambda\mu\nu(\Delta f)\,du\,dv\,dw:$$

$$\Delta f = \frac{1}{\lambda\mu\nu}\left[\frac{\partial}{\partial u}\left(\frac{\mu\nu}{\lambda}\frac{\partial f}{\partial u}\right) + \frac{\partial}{\partial v}\left(\frac{\lambda\nu}{\mu}\frac{\partial f}{\partial v}\right) + \frac{\partial}{\partial w}\left(\frac{\lambda\mu}{\nu}\frac{\partial f}{\partial w}\right)\right].$$

Let us apply this to spherical coordinates r, ϕ, θ:

$$\begin{cases} x = r\sin\phi\cos\theta \\ y = r\sin\phi\sin\theta \\ z = r\cos\phi. \end{cases}$$

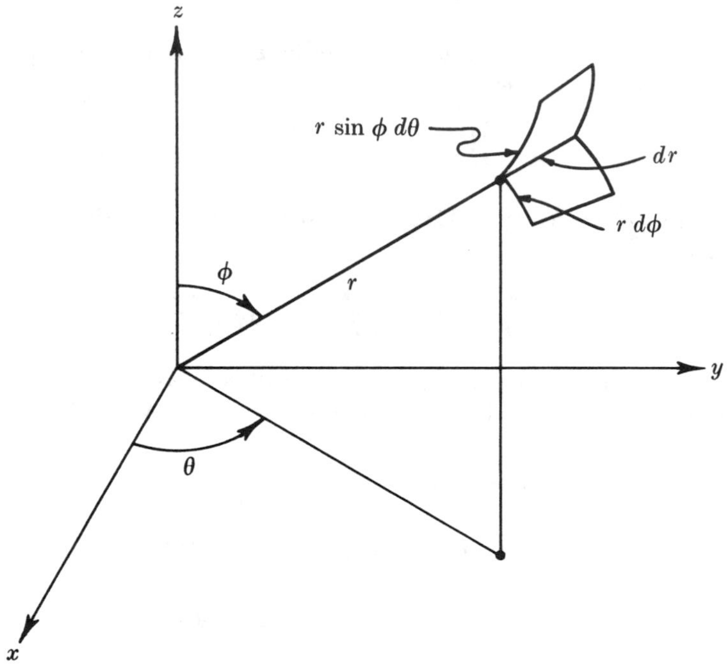

The orthogonality is easily checked (it is obvious geometrically) and we have

$$\sigma_1 = dr, \qquad \sigma_2 = r\,d\phi, \qquad \sigma_3 = r\sin\phi\,d\theta,$$

$$\Delta f = \frac{1}{r^2\sin\phi}\left[\frac{\partial}{\partial r}\left(r^2\sin\phi\,\frac{\partial f}{\partial r}\right) + \frac{\partial}{\partial\phi}\left(\sin\phi\,\frac{\partial f}{\partial\phi}\right) + \frac{\partial}{\partial\theta}\left(\frac{1}{\sin\phi}\,\frac{\partial f}{\partial\theta}\right)\right].$$

4.5. Surfaces

We study a smooth surface Σ in $\mathbf{E}^3$. We choose a moving frame $\mathbf{e}$ at each point $\mathbf{x}$ of Σ in such a way that $\mathbf{e}_3$ is the normal to the surface. Then $\mathbf{e}_1$ and $\mathbf{e}_2$ span the tangent plane at each point. We shall see how the equations of Section 4.1 specialize.

Since **x** is constrained to move in the surface, $d\mathbf{x}$ must lie in the tangent plane, $\sigma_3 = 0$:

$$d\mathbf{x} = \sigma_1 \mathbf{e}_1 + \sigma_2 \mathbf{e}_2 .$$

It is clear that the two-form $\sigma_1 \sigma_2$ represents the element of area of Σ.

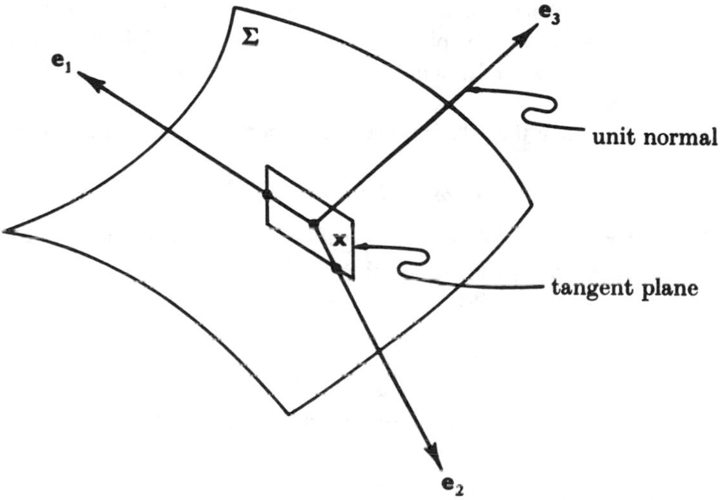

We exploit the skew-symmetry of Ω by writing

$$\Omega = \begin{pmatrix} 0 & \varpi & -\omega_1 \\ -\varpi & 0 & -\omega_2 \\ \omega_1 & \omega_2 & 0 \end{pmatrix}.$$

The structure and integrability conditions now reduce to

Structure equations

$$\left\{\begin{aligned} d\mathbf{x} &= \sigma_1 \mathbf{e}_1 + \sigma_2 \mathbf{e}_2 \\ d\mathbf{e}_1 &= \varpi \mathbf{e}_2 - \omega_1 \mathbf{e}_3 \\ d\mathbf{e}_2 &= -\varpi \mathbf{e}_1 - \omega_2 \mathbf{e}_3 \\ d\mathbf{e}_3 &= \omega_1 \mathbf{e}_1 + \omega_2 \mathbf{e}_2 \end{aligned}\right\}$$

Integrability conditions

$$\left\{\begin{aligned} d\sigma_1 &= \varpi \sigma_2 \\ d\sigma_2 &= -\varpi \sigma_1 \\ \sigma_1 \omega_1 + \sigma_2 \omega_2 &= 0 \\ d\varpi + \omega_1 \omega_2 &= 0 \\ d\omega_1 &= \varpi \omega_2 \\ d\omega_2 &= -\varpi \omega_1 \end{aligned}\right\}$$

In a certain sense, all of local surface theory is contained in these equations. It remains to interpret them in terms of curvatures, curves on the surface, etc. We illustrate a little of this.

As already remarked, $\sigma_1\sigma_2$ is the element of area on Σ. As $\mathbf{x}$ moves over Σ, $\mathbf{e}_3$ moves over a region on the unit sphere $\mathbf{S}^2$, called the *normal*, or *spherical, image* of Σ. Since $\mathbf{e}_1$ and $\mathbf{e}_2$ are orthogonal to $\mathbf{e}_3$, they lie in the tangent plane to the spherical image and form a frame there. We see that the equation $d\mathbf{e}_3 = \omega_1\mathbf{e}_1 + \omega_2\mathbf{e}_2$ plays the same rôle for the spherical image as $d\mathbf{x} = \sigma_1\mathbf{e}_1 + \sigma_2\mathbf{e}_2$ does for Σ, hence $\omega_1\omega_2$ represents the element of area of the spherical image.

Since there is only one linearly independent 2-form on the 2-dimensional space Σ, we have

$$\omega_1\omega_2 = K\sigma_1\sigma_2$$

where K is a scalar called the *Gaussian curvature*. We shall see shortly that it is entirely independent of the choice of $\mathbf{e}_1$ and $\mathbf{e}_2$.

Similarly $\sigma_1\omega_2 - \sigma_2\omega_1$ is a 2-form on Σ, and so

$$\sigma_1\omega_2 - \sigma_2\omega_1 = 2H\sigma_1\sigma_2$$

defines a scalar H called the *mean curvature* of Σ.

The one-forms ω_1, ω_2 are linear combinations of σ_1 and σ_2. Because of the relation

$$\sigma_1\omega_1 + \sigma_2\omega_2 = 0$$

we have a symmetry in the coefficients:

$$\omega_1 = p\sigma_1 + q\sigma_2$$
$$\omega_2 = q\sigma_1 + r\sigma_2.$$

We easily have from this

$$2H = p + r, \qquad K = pr - q^2.$$

The characteristic roots of the symmetric matrix

$$\begin{pmatrix} p & q \\ q & r \end{pmatrix}$$

are called the *principal curvatures* κ_1, κ_2 of Σ. We consequently have

$$2H = \kappa_1 + \kappa_2, \qquad K = \kappa_1\kappa_2.$$

From the relation $d\varpi + \omega_1\omega_2 = 0$ we have

$$d\varpi + K\sigma_1\sigma_2 = 0.$$

This relation gives us K once we know σ_1, σ_2 and ϖ. But the relations

$$d\sigma_1 = \varpi\sigma_2, \qquad d\sigma_2 = -\varpi\sigma_1$$

suffice to determine ϖ once σ_1 and σ_2 are given. (For then $d\sigma_1 = a\sigma_1\sigma_2$ and $d\sigma_2 = b\sigma_1\sigma_2$ are determined and we must have $\varpi = a\sigma_1 + b\sigma_2$.) In total then, *$K$ is completely determined analytically from σ_1 and σ_2.* This contains the theorem of Gauss that the curvature K is an intrinsic invariant of Σ, independent of how Σ is imbedded in $\mathbf{E}^3$, so long as the distance between points of Σ measured along Σ (on geodesics, or shortest paths) is preserved locally.

When we apply vector operations to vectors with differential form coefficients, we must always combine the coefficients according to the rules of exterior algebra and pay strict attention to the ordering of the factors. With this we form vector (cross) products:

$$\begin{aligned} d\mathbf{x} \times d\mathbf{x} &= (\sigma_1\mathbf{e}_1 + \sigma_2\mathbf{e}_2) \times (\sigma_1\mathbf{e}_1 + \sigma_2\mathbf{e}_2) \\ &= \sigma_1{}^2(\mathbf{e}_1 \times \mathbf{e}_1) + \sigma_2{}^2(\mathbf{e}_2 \times \mathbf{e}_2) + \sigma_1\sigma_2(\mathbf{e}_1 \times \mathbf{e}_2) + \sigma_2\sigma_1(\mathbf{e}_2 \times \mathbf{e}_1). \end{aligned}$$

Now $\sigma_1{}^2 = 0$ (and $\mathbf{e}_1 \times \mathbf{e}_1 = 0$), etc. Also

$$\begin{aligned} \sigma_2\sigma_1(\mathbf{e}_2 \times \mathbf{e}_1) &= (-\sigma_1\sigma_2)(-\mathbf{e}_1 \times \mathbf{e}_2) \\ &= (\sigma_1\sigma_2)\,\mathbf{e}_3, \end{aligned}$$

so finally

$$d\mathbf{x} \times d\mathbf{x} = 2(\sigma_1\sigma_2)\,\mathbf{e}_3$$

and we have obtained the *vectorial area element*.
Precisely, the vectorial area element is

$$(\sigma_1\sigma_2)\,\mathbf{e}_3,$$

a vector directed along the normal with magnitude $\sigma_1\sigma_2$, the element of area of Σ. Since

$$\begin{aligned} d\mathbf{x} \times d\mathbf{x} &= (dx, dy, dz) \times (dx, dy, dz) \\ &= 2(dy\,dz, dz\,dx, dx\,dy) \end{aligned}$$

we have

$$(dy\,dz, dz\,dx, dx\,dy) = (\sigma_1\sigma_2)\,\mathbf{e}_3.$$

If $\mathbf{v} = (P, Q, R)$ is a vector field, then

$$\begin{aligned} \int_\Sigma (P\,dy\,dz + Q\,dz\,dx + R\,dx\,dy) &= \int \mathbf{v}\cdot(\sigma_1\sigma_2\mathbf{e}_3) \\ &= \int_\Sigma (\mathbf{v}\cdot\mathbf{e}_3)(\sigma_1\sigma_2) \end{aligned}$$

is the *flux* of $\mathbf{v}$ through Σ.

Similarly we have

$$\begin{aligned} d\mathbf{x} \times d\mathbf{x} &= 2(\sigma_1\sigma_2)\,\mathbf{e}_3 \\ d\mathbf{x} \times d\mathbf{e}_3 &= 2H(\sigma_1\sigma_2)\,\mathbf{e}_3 \\ d\mathbf{e}_3 \times d\mathbf{e}_3 &= 2K(\sigma_1\sigma_2)\,\mathbf{e}_3 \end{aligned}$$

which shows the independence of H and K on the tangent vectors $\mathbf{e}_1$, $\mathbf{e}_2$.

If f is a function on Σ with

$$df = a_1\sigma_1 + a_2\sigma_2,$$

then on Σ,

$$*df = -a_2\sigma_1 + a_1\sigma_2,$$

$$d*df = d(-a_2\sigma_1 + a_1\sigma_2) = (\Delta f)\sigma_1\sigma_2$$

defines the *Laplacian of f on the surface* or the *second Beltrami operator* Δ. The same works for vectors and we have

$$d\mathbf{x} = \sigma_1\mathbf{e}_1 + \sigma_2\mathbf{e}_2,$$

$$*d\mathbf{x} = \sigma_2\mathbf{e}_1 - \sigma_1\mathbf{e}_2.$$

We notice that

$$d\mathbf{x} \times \mathbf{e}_3 = (\sigma_1\mathbf{e}_1 + \sigma_2\mathbf{e}_2) \times \mathbf{e}_3$$

$$= \sigma_2\mathbf{e}_1 - \sigma_1\mathbf{e}_2,$$

hence

$$*d\mathbf{x} = d\mathbf{x} \times \mathbf{e}_3,$$

$$d*d\mathbf{x} = -d\mathbf{x} \times d\mathbf{e}_3 = -2H(\sigma_1\sigma_2)\,\mathbf{e}_3$$

and so

$$\Delta\mathbf{x} = (\Delta x, \Delta y, \Delta z) = -2H\mathbf{e}_3.$$

A *minimal surface* (surface of stationary area) is one for which the mean curvature vanishes, $H = 0$. We have proved: *The coordinate functions x, y, z are harmonic on each minimal surface.* (That is, they satisfy $\Delta x = \Delta y = \Delta z = 0$.)

In this section we have given a sample of how the exterior calculus fits into the classical differential geometry of surfaces. Further material will be found in Sections 8.1 and 8.2, but there is much of the subject that we cannot cover in this text. A treatment from this point of view of exterior calculus which is not quite completely satisfactory and which unfortunately is embellished with historical comments often in bad taste is found in Blaschke [3].

4.6. Maxwell's Field Equations

In classical electromagnetic field theory one deals with the following quantities:

$\mathbf{E}$ = electric field

$\mathbf{B}$ = magnetic induction

$\mathbf{D}$ = dielectric displacement

$\mathbf{H}$ = magnetic field

$\mathbf{J}$ = electric current density

ρ = charge density.

These are all functions of the space variables x^1, x^2, x^3 and the time t. The basic Maxwell equations in ordinary vector language are

(i) $\operatorname{curl} \mathbf{E} = -\dfrac{1}{c}\dfrac{\partial \mathbf{B}}{\partial t}$ (Faraday's law of induction)

(ii) $\operatorname{curl} \mathbf{H} = \dfrac{4\pi}{c}\mathbf{J} + \dfrac{1}{c}\dfrac{\partial \mathbf{D}}{\partial t}$ (Ampère's law)

(iii) $\operatorname{div} \mathbf{D} = 4\pi\rho$ (continuity)

(iv) $\operatorname{div} \mathbf{B} = 0$ (nonexistence of true magnetism)

Here c is the speed of light. We shall put these equations into the language of exterior forms. To this end, we set

$$\begin{aligned}
\alpha &= (E_1\, dx^1 + E_2\, dx^2 + E_3\, dx^3)(c\, dt) \\
&\quad + (B_1\, dx^2\, dx^3 + B_2\, dx^3\, dx^1 + B_3\, dx^1\, dx^2), \\
\beta &= -(H_1\, dx^1 + H_2\, dx^2 + H_3\, dx^3)(c\, dt) \\
&\quad + (D_1\, dx^2\, dx^3 + D_2\, dx^3\, dx^1 + D_3\, dx^1\, dx^2), \\
\gamma &= (J_1\, dx^2\, dx^3 + J_2\, dx^3\, dx^1 + J_3\, dx^1\, dx^2)\, dt - \rho\, dx^1\, dx^2\, dx^3.
\end{aligned}$$

Equations (i) and (iv) become

$$d\alpha = 0.$$

Equations (ii) and (iii) become

$$d\beta + 4\pi\gamma = 0.$$

Applying d to this last equation yields

$$d\gamma = 0,$$

in vector notation

$$\operatorname{div} \mathbf{J} + \frac{\partial \rho}{\partial t} = 0.$$

From the equation $d\alpha = 0$ one concludes, at least in any region of space-time which can be shrunken to a point, that there is a one-form λ such that

$$d\lambda = \alpha.$$

We introduce the *vector* potential $\mathbf{A}$ and a scalar A_0 by writing

$$\lambda = A_1\, dx^1 + A_2\, dx^2 + A_3\, dx^3 + A_0 c\, dt.$$

The equation $d\lambda = \alpha$ in vector form is

$$\begin{cases} \operatorname{curl} \mathbf{A} = \mathbf{B} \\[4pt] \operatorname{grad} A_0 - \dfrac{1}{c}\dfrac{\partial \mathbf{A}}{\partial t} = \mathbf{E}. \end{cases}$$

In *free space*, everything simplifies according to

$$\mathbf{E} = \mathbf{D}, \qquad \mathbf{H} = \mathbf{B},$$
$$\mathbf{J} = 0, \qquad \rho = 0$$

so that the Maxwell equations become

$$\begin{cases} \operatorname{curl} \mathbf{E} = -\dfrac{1}{c}\dfrac{\partial \mathbf{H}}{\partial t} & \operatorname{div} \mathbf{E} = 0 \\[2ex] \operatorname{curl} \mathbf{H} = \dfrac{1}{c}\dfrac{\partial \mathbf{E}}{\partial t} & \operatorname{div} \mathbf{H} = 0. \end{cases}$$

We introduce the *Lorentz metric* into 4-space whereby

$$dx^1,\ dx^2,\ dx^3,\ c\,dt$$

is an orthonormal basis:

$$(dx^i, dx^j) = \delta^{ij}, \qquad (dx^i, c\,dt) = 0,$$
$$(c\,dt, c\,dt) = -1.$$

The signature is $3 - 1 = 2$.

According to the formulas of Section 2.7,

$$*(dx^1\,dx^2) = -dx^3(c\,dt), \qquad \text{etc.},$$
$$*(dx^1\,c\,dt) = dx^2\,dx^3, \qquad \text{etc.}$$

We see that

$$\alpha = (E_1\,dx^1 + \cdots)(c\,dt) + (H_1\,dx^2\,dx^3 + \cdots),$$
$$\beta = -(H_1\,dx^1 + \cdots)(c\,dt) + (E_1\,dx^2\,dx^3 + \cdots)$$
$$= *\alpha.$$

Consequently Maxwell's equations in free space are simply

$$\begin{cases} d\alpha = 0 \\ d*\alpha = 0. \end{cases}$$

We return to the general situation and refine our analysis by introducing one-forms:

$$\omega_1 = E_1\,dx^1 + E_2\,dx^2 + E_3\,dx^3$$
$$\omega_2 = B_1\,dx^2\,dx^3 + B_2\,dx^3\,dx^1 + B_3\,dx^1\,dx^2$$
$$\omega_3 = H_1\,dx^1 + H_2\,dx^2 + H_3\,dx^3$$
$$\omega_4 = D_1\,dx^2\,dx^3 + D_2\,dx^3\,dx^1 + D_3\,dx^1\,dx^2$$
$$\omega_5 = J_1\,dx^2\,dx^3 + J_2\,dx^3\,dx^1 + J_3\,dx^1\,dx^2.$$

These involve space variable differentials only. Now we interpret d' to denote the *exterior derivative with respect to space variables only.* We introduce $\partial/\partial t$ in this form

$$\frac{\partial}{\partial t}(\omega_1) = \dot{\omega}_1 = \dot{E}_1\,dx^1 + \cdots, \text{etc.}$$

Now the Maxwell equations are

$$\begin{cases} d'\omega_1 = -\dfrac{1}{c}\,\dot{\omega}_2 \\ d'\omega_3 = \dfrac{4\pi}{c}\,\omega_5 + \dfrac{1}{c}\,\dot{\omega}_4 \\ d'\omega_2 = 0 \\ d'\omega_4 = 4\pi\rho\,dx^1\,dx^2\,dx^3. \end{cases}$$

The *Poynting energy-flux vector* $\mathbf{S}$ is introduced by

$$\mathbf{S} = \left(\frac{c}{4\pi}\right)\mathbf{E}\times\mathbf{H}$$

that is

$$\left(\frac{c}{4\pi}\right)\omega_1\wedge\omega_3 = S_1\,dx^2\,dx^3 + S_2\,dx^3\,dx^1 + S_3\,dx^1\,dx^2.$$

Poynting's theorem,

$$\left(\frac{1}{4\pi}\right)\dot{\mathbf{B}}\cdot\mathbf{H} + \mathbf{E}\cdot\mathbf{J} + \left(\frac{1}{4\pi}\right)\mathbf{E}\cdot\dot{\mathbf{D}} + \operatorname{div}\mathbf{S} = 0,$$

follows from

$$\begin{aligned} d'(\omega_1\wedge\omega_3) &= d'\omega_1\wedge\omega_3 - \omega_1\wedge d'\omega_3 \\ &= \left(-\frac{1}{c}\,\dot{\omega}_2\right)\wedge\omega_3 - \omega_1\wedge\left(\frac{4\pi}{c}\,\omega_5 + \frac{1}{c}\,\dot{\omega}_4\right) \\ &= -\frac{1}{c}\,\dot{\omega}_2\wedge\omega_3 - \frac{4\pi}{c}\,\omega_1\wedge\omega_5 - \frac{1}{c}\,\omega_1\wedge\dot{\omega}_4. \end{aligned}$$

For bodies at rest, one assumes $\mathbf{D} = \kappa\mathbf{E}$, $\mathbf{B} = \mu\mathbf{H}$ where the *dielectric constant* κ and the *permeability* μ are constant in time. Then Poynting's theorem becomes

$$-\frac{\partial u}{\partial t} = \operatorname{div}\mathbf{S} + \mathbf{E}\cdot\mathbf{J}$$

where

$$u = \frac{1}{8\pi}(\kappa\mathbf{E}^2 + \mu\mathbf{H}^2)$$

is the *energy density* of the field. The quantity $\mathbf{E} \cdot \mathbf{J}$ is called the *thermo-chemical activity*.

4.7. Problems

1. Develop the formula for the Laplacian in cylindrical coordinates.

2. A complex matrix A is *unitary* if $AA^* = I$, where $A^* = {}^t\bar{A}$, the transpose conjugate of A. We call A skew-hermitian if $A^* + A = 0$. Discuss the connection between unitary and skew-hermitian matrices.

3. Show that e^A is orthogonal if A is skew-symmetric. Here

$$e^A = I + \sum_{n=1}^{\infty} \frac{A^n}{n!}$$

for the real matrix A.

4. Set up a frame and the structure equations for a sphere of radius R. Compute the curvatures.

5. Find Gaussian curvature of the surface of revolution obtained by revolving the curve

$$\begin{cases} x = \cos\theta + \ln\tan(\theta/2) \\ y = \sin\theta \\ \dfrac{\pi}{2} < \theta < \pi \end{cases}$$

about the x-axis.

6. Given a surface in the form $z = f(x, y)$, develop formulas for H and K in terms of f and its partial derivatives.

7. Let Σ be a surface. Let $\mathbf{e}_1$, $\mathbf{e}_2$ and $\mathbf{e}_1'$, $\mathbf{e}_2'$ be two moving frames of tangent vectors to Σ. Determine the relation between the corresponding ϖ and ϖ' and verify that $d\varpi = d\varpi'$.

8. Let $\mathbf{e}_1$, $\mathbf{e}_2$, $\mathbf{e}_3$ and $\mathbf{e}_1'$, $\mathbf{e}_2'$, $\mathbf{e}_3'$ be two moving frames in $\mathbf{E}^3$. Set up the orthogonal matrix relating these frames and determine how the corresponding Ω and Ω' are related.

V

Manifolds and Integration

5.1. Introduction

An n-dimensional manifold is a space which is not necessarily a Euclidean space nor is it a domain in a Euclidean space, but which, from the viewpoint of a short-sighted observer living in the space, looks just like such a domain of Euclidean space. A case in point is the two-sphere $\mathbf{S}^2$. This cannot be considered a part of the Euclidean plane $\mathbf{E}^2$. However our observer on $\mathbf{S}^2$ sees that he can describe his immediate vicinity by two coordinates and so he fails to distinguish between this and a small domain on $\mathbf{E}^2$.

We have the technical problem of describing an n-manifold with sufficient precision so that we can define functions, tensors, and differential forms on such a space. The definition which follows is motivated in this way. Each observer on the manifold has an immediate neighborhood (local coordinate neighborhood) described by n coordinates. Each point of the space must lie in at least one of these observed neighborhoods. Now if we consider simultaneously two observers, their immediate neighborhoods may overlap, and we must specify what happens in each such overlap. In the next three sections we go over these matters with some care.

After this is accomplished we tackle the problem of defining the integral of a differential form. In Sections 5 and 6 we lay the groundwork by defining chains, the geometrical sets over which forms are integrated, and in Section 7 we define the integral.

5.2. Manifolds

An n-dimensional manifold consists of a space $\mathbf{M}$ together with a collection of *local coordinate neighborhoods* $\mathbf{U}_1, \mathbf{U}_2, \cdots$ such that each point of $\mathbf{M}$ lies in at least one of these $\mathbf{U}$. On each $\mathbf{U}$ is given a coordinate system

$$x^1, \cdots, x^n$$

so that the *values* of the coordinates

$$(x^1(P), \cdots, x^n(P)),$$

where P ranges over $\mathbf{U}$, make up an open domain in Euclidean n-space $\mathbf{E}^n$.

Suppose that **U** with coordinate system

$$x^1, \cdots, x^n$$

and **V** with coordinate system

$$y^1, \cdots, y^n$$

overlap (intersect). We may express the **V** coordinates **y** of a point P in terms of the **U** coordinates **x** of this point:

$$y^i = y^i(x^1, \cdots, x^n) \qquad (i = 1, \cdots, n).$$

As part of the definition, *we assume that these functions are smooth* (differentiable as often as we please).

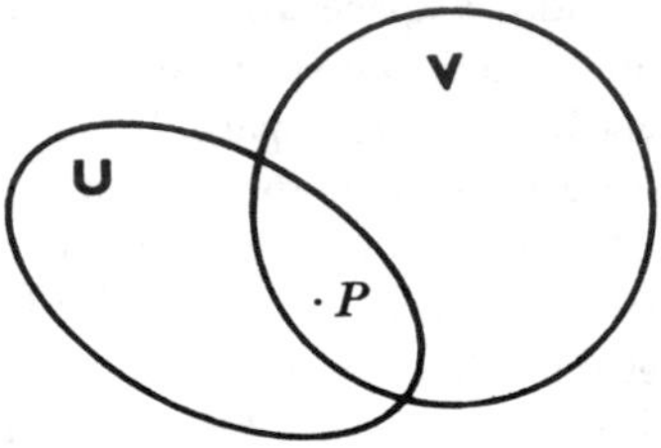

Having this formal definition out of the way, we explore some consequences. First of all, on the overlap of **U** and **V** above we may interchange the rôles of **U** and **V** to write smooth functions

$$x^j = x^j(y^1, \cdots, y^n) \qquad (j = 1, \cdots, n).$$

Substituting yields

$$y^i = y^i\left(x^1(\mathbf{y}), \cdots, x^n(\mathbf{y})\right)$$

and we may differentiate by the chain rule:

$$\delta^i_k = \sum \frac{\partial y^i}{\partial x^j} \frac{\partial x^j}{\partial y^k},$$

which has the matrix interpretation

$$\left\| \frac{\partial y^i}{\partial x^j} \right\| \cdot \left\| \frac{\partial x^j}{\partial y^k} \right\| = I.$$

We take determinants by the product rule:

$$\frac{\partial(y^1, \cdots, y^n)}{\partial(x^1, \cdots, x^n)} \cdot \frac{\partial(x^1, \cdots, x^n)}{\partial(y^1, \cdots, y^n)} = 1.$$

It follows that the Jacobian

$$\frac{\partial(y^1, \cdots, y^n)}{\partial(x^1, \cdots, x^n)} \neq 0;$$

it is different from 0 at *each* point.

A manifold is called *orientable* (two-sided) if it is possible to choose the local coordinates in the first place so that each such Jacobian (on an overlap of local coordinate neighborhoods) is positive.

Example. We make the two-sphere $\mathbf{S}^2$ into a manifold by using six coordinate neighborhoods. We set

$$\mathbf{S}^2 = \{(x, y, z) \quad \text{where} \quad x^2 + y^2 + z^2 = 1\}.$$

The neighborhoods are

$$\begin{aligned}
\mathbf{U}_1^+ &= \{x > 0\}, && \text{coordinate system } y, z.\\
\mathbf{U}_1^- &= \{x < 0\}, && \text{coordinate system } z, y.\\
\mathbf{U}_2^+ &= \{y > 0\}, && \text{coordinate system } z, x.\\
\mathbf{U}_2^- &= \{y < 0\}, && \text{coordinate system } x, z.\\
\mathbf{U}_3^+ &= \{z > 0\}, && \text{coordinate system } x, y.\\
\mathbf{U}_3^- &= \{z < 0\}, && \text{coordinate system } y, x.
\end{aligned}$$

In comparing the overlap of two of these, we shall not be pedantic and introduce different letters, hoping the reader will forgive this sloppy notation.

On the intersection of $\mathbf{U}_1^+$ and $\mathbf{U}_2^+$ we have the coordinate transformation

$$\begin{cases} y = \sqrt{1 - z^2 - x^2} \\ z = z, \qquad\qquad x > 0, \quad y > 0 \end{cases}$$

and so

$$\frac{\partial(y, z)}{\partial(z, x)} = \begin{vmatrix} \dfrac{-z}{\sqrt{\ }} & \dfrac{-x}{\sqrt{\ }} \\ 1 & 0 \end{vmatrix} = \frac{x}{\sqrt{\ }} > 0.$$

On the intersection of $\mathbf{U}_1^+$ and $\mathbf{U}_3^-$,

$$\begin{cases} y = y \\ z = -\sqrt{1 - y^2 - x^2}, \qquad x > 0, \quad z < 0, \end{cases}$$

$$\frac{\partial(y, z)}{\partial(y, x)} = \begin{vmatrix} 1 & 0 \\ \dfrac{y}{\sqrt{\ }} & \dfrac{x}{\sqrt{\ }} \end{vmatrix} = \frac{x}{\sqrt{\ }} > 0.$$

On the intersection of $\mathbf{U}_2^-$ and $\mathbf{U}_3^-$,

$$\begin{cases} x = x \\ z = -\sqrt{1 - y^2 - x^2}, \qquad y < 0, \quad z < 0, \end{cases}$$

$$\frac{\partial(x, z)}{\partial(y, x)} = \begin{vmatrix} 0 & 1 \\ \dfrac{y}{\sqrt{\ }} & \dfrac{x}{\sqrt{\ }} \end{vmatrix} = -\frac{y}{\sqrt{\ }} > 0, \qquad \text{etc.}$$

Thus the two-sphere is a two-manifold, and our choice of local coordinates proves it to be orientable.

One could also cover the sphere $\mathbf{S}^2$ with a system of only two local coordinate neighborhoods by taking two opposite hemispheres, each extended slightly to make open overlapping neighborhoods.

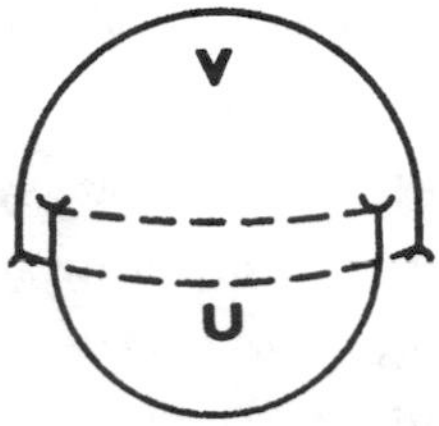

The sphere $\mathbf{S}^2$ has two opposite *orientations* (outward or inward normal, corresponding to counterclockwise or clockwise sense of rotation). Similarly an orientable n-manifold has two opposite orientations. A definite one of these is determined by the order in which local coordinates $x^1, \cdots, x^n$ are given, *up to an even permutation of this order*. Making an *odd* permutation of local coordinates gives the opposite orientation.

Let $\mathbf{M}$ be an n-manifold. To say that a real-valued function f on $\mathbf{M}$ is *smooth* at a point P of $\mathbf{M}$ means the following. Let $\mathbf{U}$ be a local coordinate neighborhood containing P with coordinates $x^1, \cdots, x^n$. We require that $f(x^1, \cdots, x^n)$ be smooth near P. This restriction on f is *independent* of the particular $\mathbf{U}$ one chooses, since two coordinate systems whose neighborhoods overlap on a region including P are themselves related by smooth functions (from the definition of manifold). A real-valued function f is *smooth* on $\mathbf{M}$ if it is smooth at each point of $\mathbf{M}$.

Similarly, if $\mathbf{M}$ and $\mathbf{N}$ are manifolds of dimensions m and n, respectively, one defines a *smooth mapping*

$$\phi\colon \quad \mathbf{M} \longrightarrow \mathbf{N}$$

by the requirements that in local coordinates $x^1, \cdots, x^m$ on $\mathbf{U}$ in $\mathbf{M}$ and $y^1, \cdots, y^n$ on $\mathbf{V}$ in $\mathbf{N}$, we have ϕ represented by smooth functions

$$y^i = y^i(x^1, \cdots, x^m) \qquad (i = 1, \cdots, n)$$

on that part of $\mathbf{U}$ which ϕ maps into $\mathbf{V}$.

A manifold $\mathbf{M}$ is called a *submanifold* of a manifold $\mathbf{N}$ provided there is a one-to-one smooth mapping

$$j\colon \quad \mathbf{M} \longrightarrow \mathbf{N}$$

which has this *regularity* property: in local coordinates (as written above), the matrix

$$\left\| \frac{\partial y^i}{\partial x^j} \right\|$$

has (maximal) rank m at each point. We refer to j itself as an *injection* or *imbedding* of $\mathbf{M}$ in $\mathbf{N}$.

This applies in particular when $\mathbf{N} = \mathbf{E}^n$ so that we may refer to submanifolds of Euclidean spaces. It is an established result of manifold theory that each m-dimensional manifold which is not too large may be imbedded in $\mathbf{E}^n$ with $n = 2m + 1$.

5.3. Tangent Vectors

We study a manifold $\mathbf{M}$ and a point P on $\mathbf{M}$. Our job is to define the tangent space at P, an n-dimensional vector space whose elements are the tangent vectors at P. Because we are not within the simple terrain of Euclidean space we cannot merely draw arrows emanating at P. We need a way of considering ordinary Euclidean vectors which depends in no way on arrows, or directed line segments. The answer is simple. We may identify Euclidean vectors with directional differentiations. Thus in case P is a point of $\mathbf{E}^3$ and $\mathbf{v} = (a, b, c)$ is a vector at P, we may identify $\mathbf{v}$ with the operator

$$\left(a\frac{\partial}{\partial x} + b\frac{\partial}{\partial y} + c\frac{\partial}{\partial z}\right)\bigg|_P .$$

This does the usual things to sums and products, which motivates the following definition.

First some notation. If $\mathbf{M}$ is a manifold, we denote by

$$\mathbf{F}^0(\mathbf{M})$$

the space of all smooth real-valued functions on $\mathbf{M}$.

Let P be a point on a manifold $\mathbf{M}$. A *tangent vector* $\mathbf{v}$ at P is an operator

$$\mathbf{v}\colon\quad \mathbf{F}^0(\mathbf{M}) \longrightarrow \mathbf{R}, \text{ the reals}$$

satisfying

(i) $\mathbf{v}(af + bg) = a\mathbf{v}(f) + b\mathbf{v}(g)$, $\quad a, b$ constant.
(ii) $\mathbf{v}(f\cdot g) = g(P)\cdot\mathbf{v}(f) + f(P)\cdot\mathbf{v}(g)$.

Thus $\mathbf{v}$ assigns to each smooth function f on $\mathbf{M}$ a real number $\mathbf{v}(f)$.

We shall first observe that if we take a constant function c, then $\mathbf{v}(c) = 0$. For setting $f = g = 0$ in (i) yields $\mathbf{v}(0) = 0$, setting $f = g = 1$ in (ii) yields $\mathbf{v}(1) = 0$, and setting $f = 1$, $a = 0$, and $g = 0$ in (i) yields $\mathbf{v}(c) = 0$. Observe that

$$\mathbf{v}(cf) = c\mathbf{v}(f)$$

for any f and constant c.

Next, suppose $x^1, \cdots, x^n$ is a local coordinate system, valid in some neighborhood of P. Then each of the operators

$$\mathbf{v}_i = \frac{\partial}{\partial x^i}\bigg|_P$$

(the vertical bar means "evaluated at P") is a tangent vector, as one easily verifies.

The totality of tangent vectors at P makes up a linear space, $\mathbf{T}_P$, called the tangent space to $\mathbf{M}$ at P. We shall show that *these vectors* $\mathbf{v}_i$ *form a basis of this tangent space.* We set

$$(x^1, \cdots, x^n)|_P = (c^1, \cdots, c^n).$$

If $\mathbf{v}$ is any tangent vector at P, set

$$\mathbf{v}(x^i) = \mathbf{v}(x^i - c^i) = a^i.$$

Now if f is any smooth function on $\mathbf{M}$, we expand f in a Taylor series up to first-order terms with the integral form of remainder:

$$f(\mathbf{x}) = f(\mathbf{c}) + \sum (x^i - c^i)\, g_i(\mathbf{x}),$$

$$g_i(\mathbf{c}) = \left.\frac{\partial f}{\partial x^i}\right|_P .$$

Then

$$\begin{aligned}\mathbf{v}(f) &= \mathbf{v}[f(\mathbf{c})] + \sum g_i(\mathbf{c})\, \mathbf{v}(x^i - c^i) + \sum (c^i - c^i)\, \mathbf{v}(g_i)\\ &= 0 + \sum a^i \left.\frac{\partial f}{\partial x^i}\right|_P + 0 = \sum a^i \left.\frac{\partial f}{\partial x^i}\right|_P ,\end{aligned}$$

hence

$$\mathbf{v} = \sum a^i \left.\frac{\partial}{\partial x^i}\right|_P$$

which establishes the result. We refer to $a^1, \cdots, a^n$ as the *components* of $\mathbf{v}$ with respect to the coordinate system $\mathbf{x}$. If $\mathbf{y}$ is another coordinate system valid at P, and

$$\mathbf{v} = \sum b^i \left.\frac{\partial}{\partial y^i}\right|_P$$

we find, by the chain rule,

$$b^i = \sum a^j \left.\frac{\partial y^i}{\partial x^j}\right|_P ,$$

the usual transformation law for contravariant components of a vector. Note here that we are working at a single point so that $\mathbf{a}$ and $\mathbf{b}$ are constant.

A *vector field* on $\mathbf{M}$ consists of a smooth assignment of a tangent vector to each point of $\mathbf{M}$. In local coordinates,

$$\mathbf{v} = \sum a^i(\mathbf{x}) \frac{\partial}{\partial x^i}, \qquad a^i(\mathbf{x}) \text{ smooth.}$$

On an overlap,

$$\mathbf{v} = \sum b^i(\mathbf{y}) \frac{\partial}{\partial y^i},$$

$$b^i(\mathbf{y}(\mathbf{x})) = \sum a^j(\mathbf{x}) \frac{\partial y^i}{\partial x^j} .$$

5.4. Differential Forms

The smooth functions on **M** will also be called 0-forms. They form a space $\mathbf{F}^0(\mathbf{M})$, the space of forms of degree 0 on **M**.

We now define a one-form at a point P of **M**. We must have an expression

$$\sum a_i\,dx^i, \qquad a_i \text{ constant}$$

for each local coordinate system (x^i) valid in a neighborhood **U** which includes P and such that any two such expressions

$$\sum a_i\,dx^i, \qquad \sum b_i\,dy^i$$

at P are related by

$$\sum b_i \left.\frac{\partial y^i}{\partial x^j}\right|_P = a_j,$$

the usual transformation law for covariant vectors. Evidently this is completely consistent with our local study in Chapter III.

Having this, we may form sums of exterior products of one-forms at P to construct p-forms at P. Now we can define a p-form on **M**. This is a smooth assignment of a p-form to each point P of **M**. If **U** is given with local coordinates (x^i), then on the neighborhood **U** we have the representation

$$\omega = \sum a_H(\mathbf{x})\,dx^H$$

with smooth functions $a_H(\mathbf{x})$ on **U**, $H = \{h_1, \cdots, h_p\}$.

If we have the representation

$$\omega = \sum b_K(\mathbf{y})\,dy^K$$

with respect to a second coordinate system which overlaps the first, then the relation between the b's and a's is given by substitution of $y^i = y^i(\mathbf{x})$ for y^i and

$$\sum \frac{\partial y^i}{\partial x^j}\,dx^j$$

for dy^i. As a consequence of our study of coordinate changes in Section 3.4, we see that the space

$$\mathbf{F}^p(\mathbf{M})$$

of p-forms on **M** is completely defined, that exterior multiplication

$$\omega \wedge \eta$$

of two-forms on **M** is accomplished by operating with one point at a time, and that the exterior derivative of a form on **M** is defined by working it out in each local coordinate system.

All the rules of Chapter III are readily verified,

$$d(\omega \wedge \eta) = d\omega \wedge \eta + (-1)^{(\deg \omega)}\, \omega \wedge d\eta$$

for example.

If $\mathbf{M}$ and $\mathbf{N}$ are two manifolds and

$$\phi\colon \quad \mathbf{M} \longrightarrow \mathbf{N}$$

is a smooth mapping, then there is a natural induced mapping ϕ^*,

$$\phi^*\colon \quad \mathbf{F}^p(\mathbf{N}) \longrightarrow \mathbf{F}^p(\mathbf{M})$$

which again is defined by applying the local construction in one local coordinate system at a time and piecing together the results. As in the local theory, we have the results

(i) $\phi^*(\omega + \eta) = \phi^*\omega + \phi^*\eta.$
(ii) $\phi^*(\lambda \wedge \mu) = (\phi^*\lambda) \wedge (\phi^*\mu).$
(iii) $d(\phi^*\omega) = \phi^*(d\omega).$

The last of these can be expressed by means of a *commutative diagram*

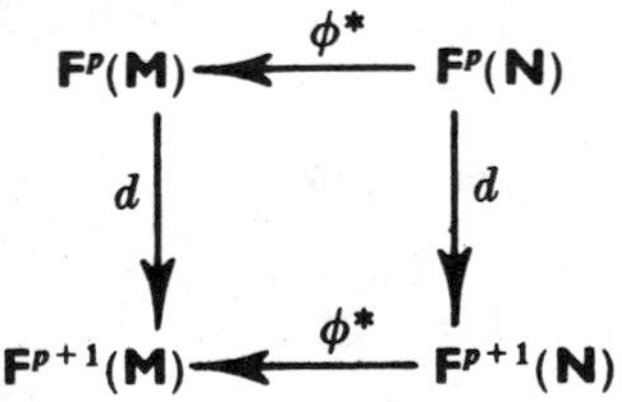

Each of the two possible paths from $\mathbf{F}^p(\mathbf{N})$ to $\mathbf{F}^{p+1}(\mathbf{M})$ leads to the same result.

In practice, one often constructs differential forms on a manifold this way. One knows in advance several smooth functions $f, g, \cdots$ on $\mathbf{M}$. From these one constructs one-forms $df, dg, \cdots$ and from these in turn forms of higher degrees by taking exterior products.

Example 1. On the two sphere $\mathbf{S}^2$ considered in Section 5.2, the functions x, y, z are smooth 0-forms. Thus dx, dy, dz are one-forms and $dx\,dy, dy\,dz$, etc., are two-forms. On the neighborhood $\mathbf{U}_1^+$ we have

$$x^2 = 1 - y^2 - z^2,$$

$$dx = \frac{-y\,dy - z\,dz}{x},$$

$$dx\,dy = \frac{-y\,dy - z\,dz}{x}\,dy = \frac{z}{x}\,dy\,dz, \qquad \text{etc.}$$

Example 2. The circle $\mathbf{S}^1 = \{x^2 + y^2 = 1\}$. Here x and y are functions on $\mathbf{S}^1$ and

$$x\,dx + y\,dy = 0.$$

This means we can define a one-form α. At a point where $x \neq 0$,

$$\alpha = \frac{dy}{x}.$$

At a point where $x = 0$, we have $y \neq 0$ and

$$\alpha = -\frac{dx}{y}.$$

On any arc of $\mathbf{S}^1$ which is not the complete circle, we can find a function θ such that

$$x = \cos\theta, \qquad y = \sin\theta,$$

hence

$$\alpha = d\theta.$$

It must be emphasized that no such function θ exists on *all* of $\mathbf{S}^1$—it would have to jump by 2π somewhere.

5.5. Euclidean Simplices

In this section we shall describe the standard building blocks which we later piece together to form fields of integration, p-dimensional spreads in a manifold over which we can integrate p-forms. These building blocks will be called Euclidean simplices of various dimensions—we shall omit repetition of the adjective Euclidean in this section, but we understand that everything takes place in Euclidean space.

A 0-*simplex* is a single point (P_0).

A 1-*simplex* is a directed closed segment on a straight line. It is completely determined by its ordered pair of vertices (P_0P_1).

A 2-*simplex* is a closed triangle with vertices taken in some definite order. It is completely determined by its ordered triple of vertices in the proper order,

$$(P_0, P_1, P_2).$$

Similarly one has a 3-*simplex* based on an ordered quadruple

$$(P_0, P_1, P_2, P_3)$$

of four points, no three collinear. Geometrically it represents a tetrahedron. Finally, an n-*simplex* is the closed convex hull

$$(P_0, \cdots, P_n)$$

of $(n + 1)$ independent† points taken in a definite order. The geometrical set so spanned consists of all points

$$P = t_0P_0 + \cdots + t_nP_n, \qquad t_i \geqq 0, \qquad \sum t_i = 1,$$

i.e., all possible centroids of systems of nonnegative masses $t_0, \cdots, t_n$ located at $P_0, \cdots, P_n$, respectively.

The *boundary* $\partial \mathbf{s}$ of a simplex $\mathbf{s}$ is a formal sum of simplices of one lower dimension with integer coefficients:

$$\partial(P_0, P_1, \cdots, P_n) = \sum_{i=0}^{n} (-1)^i (P_0, P_1, \cdots, P_{i-1}, P_{i+1}, \cdots, P_n).$$

An examination of the lower dimensional cases convinces one that this is consistent with the customary ideas on boundaries of oriented regions.

$$\partial(P_0, P_1) = (P_1) - (P_0),$$

$$\partial(P_0, P_1, P_2) = (P_1, P_2) - (P_0, P_2) + (P_0, P_1),$$

$$\partial(P_0, P_1, P_2, P_3) = (P_1, P_2, P_3) - (P_0, P_2, P_3) + (P_0, P_1, P_3) - (P_0, P_1, P_2).$$

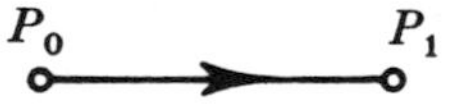

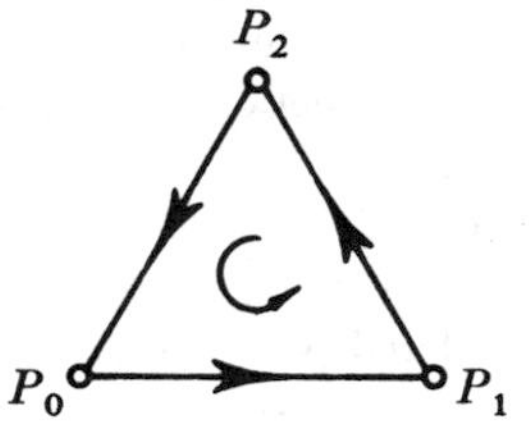

In the triangle, the ordering of the vertices gives a sense of rotation of the triangle. In the tetrahedron, the ordering of the vertices gives a right-handed screw sense in space and induces a positive sense of rotation in each triangular face (outward drawn normal). One thinks of each minus sign in $\partial \mathbf{s}$ as representing a reversal in this rotation sense. The result is that $\partial(P_0, \cdots, P_3)$ represents the oriented geometric boundary of the tetrahedron according to the outward drawn normal.

† This means that the n vectors $(P_1 - P_0)$, $(P_2 - P_0), \ldots, (P_n - P_0)$ are linearly independent.

An *n-chain* is a formal sum

$$\mathbf{c} = \sum a^i \mathbf{s}_i$$

where the a^i are constants and the $\mathbf{s}_i$ are n-simplices. Its *boundary* is defined by

$$\partial \mathbf{c} = \sum a^i (\partial \mathbf{s}_i).$$

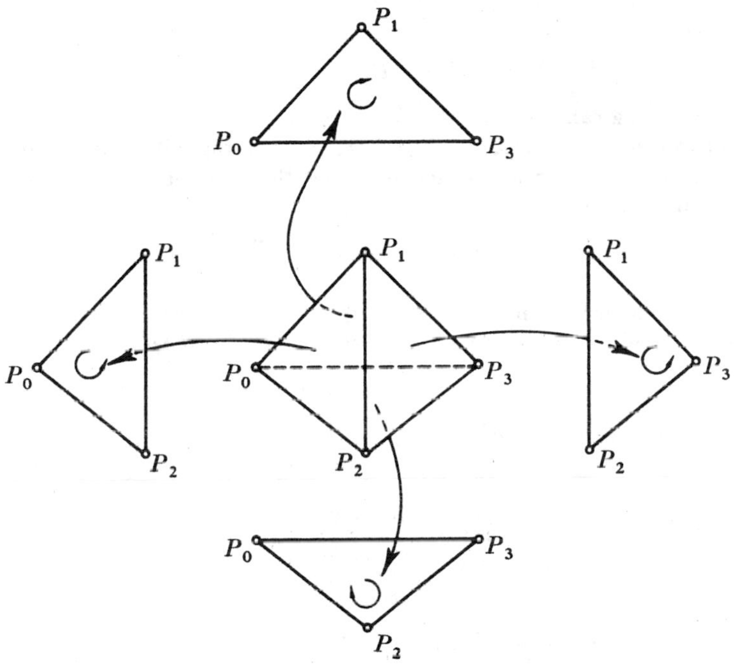

A basic result is that the boundary of each chain itself has zero boundary:

$$\partial[\partial \mathbf{c}] = 0.$$

It suffices to check this for simplices. Let us try low-dimensional cases:

$$\begin{aligned}
\partial[\partial(P_0, P_1, P_2)] &= \partial(P_1, P_2) - \partial(P_0, P_2) + \partial(P_0, P_1) \\
&= [(P_2) - (P_1)] - [(P_2) - (P_0)] + [(P_1) - (P_0)] = 0, \\
\partial[\partial(P_0, \cdots, P_3)] &= [(P_2, P_3) - (P_1, P_3) + (P_1, P_2)] \\
&\quad - [(P_2, P_3) - (P_0, P_3) + (P_0, P_2)] \\
&\quad + [(P_1, P_3) - (P_0, P_3) + (P_0, P_1)] \\
&\quad - [(P_1, P_2) - (P_0, P_2) + (P_0, P_1)] = 0,
\end{aligned}$$

which illustrates the general idea; each face occurs twice with opposite signs.

More generally, in computing

$$\partial[\partial(P_0, \cdots, P_n)],$$

one obtains

$$(P_0, \cdots, P_{i-1}, P_{i+1}, \cdots, P_{j-1}, P_{j+1}, \cdots, P_n)$$

twice, with opposite signs, once each from

$$\partial(P_0, \cdots, P_{i-1}, P_{i+1}, \cdots, P_n)$$

and

$$\partial(P_0, \cdots, P_{j-1}, P_{j+1}, \cdots, P_n),$$

so that everything cancels.

Given two n-simplices $(P_0, \cdots, P_n)$, $(Q_0, \cdots, Q_n)$, there is a unique linear correspondence between them which preserves the ordering of the vertices. It is given by

$$\sum_0^n t_i P_i \longleftrightarrow \sum_0^n t_i Q_i \qquad (t_i \geqq 0, \qquad \sum_0^n t_i = 1).$$

It is convenient for defining integrals to have standard models of the simplices of each dimension. We define the *standard n-simplex*

$$\bar{\mathbf{s}}^n = (R_0, \cdots, R_n)$$

as the simplex in $\mathbf{E}^n$ based on

$$R_0 = 0$$
$$R_1 = (10 \cdots 0)$$
$$R_2 = (010 \cdots 0)$$
$$\cdots\cdots\cdots$$
$$R_n = (00 \cdots 01).$$

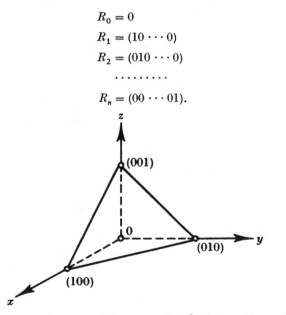

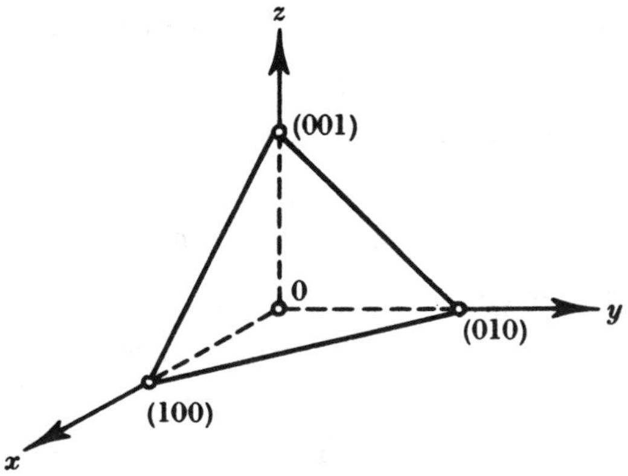

We must now agree on a certain convention for integration. Let ω be an

n-form defined on a domain $\mathbf{U}$ of $\mathbf{E}^n$ which includes $\bar{\mathbf{s}}^n$. We wish to define

$$\int_{\bar{\mathbf{s}}^n} \omega.$$

We do this by writing ω in the unique way

$$\omega = A(x^1, \cdots, x^n)\, dx^1\, dx^2 \cdots dx^n$$

with the variables in their natural order, and then setting

$$\int_{\bar{\mathbf{s}}^n} \omega = \int_{\bar{\mathbf{s}}^n} A(\mathbf{x})\, dx^1\, dx^2 \cdots dx^n,$$

where the right-hand side is now the standard ordinary n-fold integration, which may be evaluated by any scheme of iteration, *regardless of what order in which the variables are taken.*

For example, if $\omega = dz\, dy\, dx$, then

$$\int_{\bar{\mathbf{s}}^3} \omega = -\int_{\bar{\mathbf{s}}^3} dx\, dy\, dz = -\int_0^1 dy \int_0^{1-y} dx \int_0^{1-x-y} dz = -1/6.$$

5.6. Chains and Boundaries

Now we consider a manifold $\mathbf{M}$ and we shall define an *n-simplex* in $\mathbf{M}$. As a preliminary definition, this consists of three things: a Euclidean n-simplex $\mathbf{s}^n$, an n-dimensional neighborhood $\mathbf{U}$ of $\mathbf{s}^n$ in Euclidean space,† and a smooth mapping ϕ,

$$\phi\colon\ \mathbf{U} \longrightarrow \mathbf{M}.$$

We denote this preliminary simplex by

$$(\mathbf{s}^n, \mathbf{U}, \phi).$$

If we are given a second one,

$$(\mathbf{t}^n, \mathbf{V}, \psi),$$

it will be considered the same as the first provided

$$\phi\left(\sum_0^n t_i P_i\right) = \psi\left(\sum_0^n t_i Q_i\right) \qquad \left(t_i \geqq 0, \quad \sum_0^n t_i = 1\right),$$

where

$$\mathbf{s}^n = (P_0, P_1, \cdots, P_n), \qquad \mathbf{t}^n = (Q_0, Q_1, \cdots, Q_n).$$

In other words, if we set up the natural order-preserving linear equivalence

† That is, a neighborhood in the smallest flat submanifold of Euclidean space containing $\mathbf{s}^n$. This smallest flat submanifold is the totality of points $\sum_0^n t_i P_i$, t_i real, $\sum t_i = 1$, where $\mathbf{s}^n = (P_0, \ldots, P_n)$.

between $\mathbf{s}^n$ and $\mathbf{t}^n$:

$$\mathbf{s}^n \longleftrightarrow \mathbf{t}^n,$$

then $\phi(P) = \psi(Q)$ whenever P and Q are corresponding points. This is also expressed by the commutative diagram

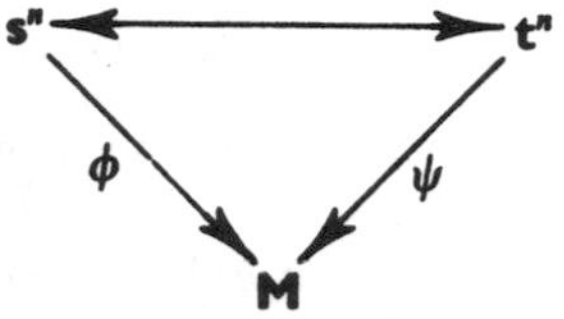

The totality of these preliminary simplices $(\mathbf{s}^n, \mathbf{U}, \phi)$ which in this way are identified with a single one make up an object which we call an *n-simplex in* $\mathbf{M}$, denoted by a symbol $\boldsymbol{\sigma}^n$.

The open neighborhoods $\mathbf{U}$ we have introduced merely serve to eliminate difficulties with differentiability on the boundary.

If $\boldsymbol{\sigma}^n$ is a simplex represented by $(\mathbf{s}^n, \mathbf{U}, \phi)$, then $\mathbf{s}^n$ has faces $\mathbf{t}_0, \cdots, \mathbf{t}_n$, each a Euclidean $(n-1)$-simplex, where

$$\partial \mathbf{s}^n = \sum \pm \mathbf{t}_i .$$

By restricting ϕ to the various $\mathbf{t}_i$, each extended a little in $\mathbf{U}$ to make open neighborhoods $\mathbf{V}_i$, we define the *faces* of $\boldsymbol{\sigma}^n$, each represented by

$$\tau_i = (\mathbf{t}_i, \mathbf{V}^i, \phi)$$

and the corresponding *boundary*

$$\partial \boldsymbol{\sigma}^n = \sum \pm \tau_i .$$

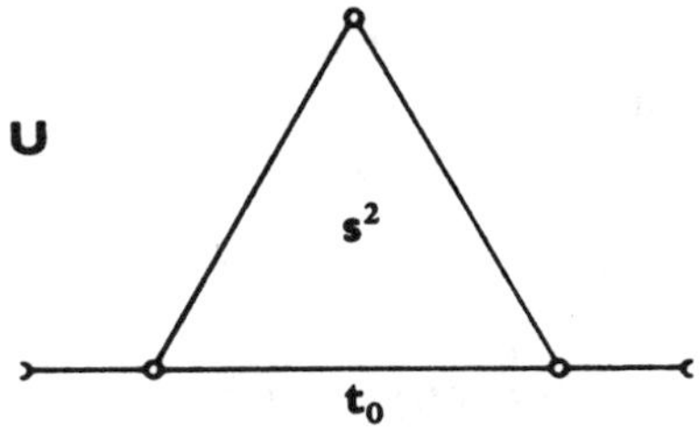

This is an $(n-1)$-chain in $\mathbf{M}$. By an *n-chain* $\mathbf{c}$ of $\mathbf{M}$ we mean a formal sum

$$\mathbf{c} = \sum_i a_i \boldsymbol{\sigma}_i^n$$

with constant coefficients a_i and n-simplices $\boldsymbol{\sigma}_i^n$. Chains may be added and multiplied by constants. We denote by

$$\mathbf{C}_n(\mathbf{M})$$

the set of all n-chains† on $\mathbf{M}$. We set

$$\partial \mathbf{c} = \sum a_i \, \partial \boldsymbol{\sigma}_i^n \qquad \text{for} \qquad \mathbf{c} = \sum a_i \boldsymbol{\sigma}_i^n .$$

Thus

$$\partial: \quad \mathbf{C}_n(\mathbf{M}) \longrightarrow \mathbf{C}_{n-1}(\mathbf{M}) \qquad (n = 1, 2, \cdots).$$

The basic property of the boundary operator ∂ follows readily from the corresponding Euclidean situation: *for each n-chain* $\mathbf{c}$,

$$\partial(\partial \mathbf{c}) = 0.$$

A *cycle* is a chain $\mathbf{z}$ whose boundary vanishes, $\partial \mathbf{z} = 0$.

A *bounding cycle* (or simply *boundary*) $\mathbf{b}$ is a chain which is the boundary of a chain of one higher dimension, $\mathbf{b} = \partial \mathbf{c}$.

Each boundary is a cycle, for if $\mathbf{b} = \partial \mathbf{c}$, then

$$\partial(\mathbf{b}) = \partial(\partial \mathbf{c}) = 0.$$

One further thing to be noted is this. In our preliminary definition of a simplex $(\mathbf{s}^n, \mathbf{U}, \phi)$ we do not require that the smooth mapping ϕ on $\mathbf{U}$ into $\mathbf{M}$ be a one-to-one mapping. Indeed, it may happen that it takes all of $\mathbf{s}^n$ into a lower dimensional space, even into a single point! A close analysis shows that not only is there no harm in allowing such "bad" mappings but that there are very great technical difficulties involved in attempting to avoid them.

5.7. Integration of Forms

Our data is a manifold $\mathbf{M}$ of any dimension, a p-form ω on $\mathbf{M}$ and a p-chain $\mathbf{c}$ on $\mathbf{M}$. We must define

$$\int_{\mathbf{c}} \omega .$$

First we set

$$\mathbf{c} = \sum a_i \boldsymbol{\sigma}_i$$

where the a_i are constants and the $\boldsymbol{\sigma}_i$ are p-simplices and write

$$\int_{\mathbf{c}} \omega = \sum a_i \int_{\boldsymbol{\sigma}_i} \omega$$

so it remains to define the integral of ω over a p-simplex $\boldsymbol{\sigma}$. Now we can represent $\boldsymbol{\sigma}$ in the form

$$(\bar{\mathbf{s}}^p, \mathbf{U}, \phi)$$

where $\bar{\mathbf{s}}^p$ is the standard p-simplex in $\mathbf{E}^p$ and ϕ is a smooth mapping of the neighborhood $\mathbf{U}$ of $\bar{\mathbf{s}}^p$ into $\mathbf{M}$. Our definition is

† Precise topological terminology: ordered singular differentiable n-chains.

$$\int_{\sigma} \omega = \int_{\bar{\mathbf{s}}^p} \phi^* \omega.$$

Since $\phi^* \omega$ is a p-form on $\mathbf{U}$, this is an ordinary p-fold integral, as discussed in the next to last section.

In application, one often does not bother to spell out in detail how a given geometrical region may be considered as a chain, but rather relies on the usual combination of experience and intuition, the latter an excellent guide in geometry. For example, suppose ω is a 2-form on $\mathbf{S}^2 = \{x^2 + y^2 + z^2 = 1\}$ and one seeks $\int \omega$ taken over $\mathbf{S}^2$. There will *usually* be a more effective procedure than using the coordinate planes to decompose the surface $\mathbf{S}^2$ into eight spherical triangles, setting up mappings of the standard triangle onto each of these, etc.

What then is the value of this rather long story on chains, boundaries, and integrals? In this age, it hardly seems necessary to defend the placing on a logical and rigorous basis things which are only understood in an intuitive sense. In addition, we have here a powerful theoretical tool as we shall see immediately in the following section on the general Stokes' theorem.

As an exercise, one could check that each of the standard tricks used to evaluate surface integrals, etc., fits into the above scheme of things. It hardly seems worth our time here.

5.8. Stokes' Theorem

The general result we establish now includes all known formulas which transform an integral into one over a one-higher dimension spread.

Let ω be a p-form on a manifold $\mathbf{M}$ and $\mathbf{c}$ a $(p + 1)$-chain. Then

$$\int_{\partial \mathbf{c}} \omega = \int_{\mathbf{c}} d\omega.$$

Since $\mathbf{c}$ is a sum of $(p + 1)$-simplices with constant coefficients, it suffices to prove

$$\int_{\partial \sigma} \omega = \int_{\sigma} d\omega$$

where σ is a $(p + 1)$-simplex. According to a representation

$$(\bar{\mathbf{s}}^{p+1}, \mathbf{U}, \phi)$$

of σ we have from the definition

$$\int_{\sigma} d\omega = \int_{\bar{\mathbf{s}}^{p+1}} \phi^* (d\omega) = \int_{\bar{\mathbf{s}}^{p+1}} d(\phi^* \omega).$$

This reduces the problem to a Euclidean one. Let η be a p-form on a neighborhood $\mathbf{U}$ of $\bar{\mathbf{s}}^{p+1}$ in $\mathbf{E}^{p+1}$. To prove

$$\int_{\partial\bar{\mathbf{s}}^{p+1}} \eta = \int_{\bar{\mathbf{s}}^{p+1}} d\eta .$$

Now

$$\eta = \sum A_i(\mathbf{x})\, dx^1 \cdots dx^{i-1}\, dx^{i+1} \cdots dx^{p+1}$$

so that it suffices to check the formula in case η is a monomial only. Since we may permute coordinates provided we are careful about signs, it suffices to take the case

$$\eta = A\, dx^1 \cdots dx^p.$$

Then

$$d\eta = (-1)^p \frac{\partial A}{\partial x^{p+1}}\, dx^1 \cdots dx^{p+1}.$$

We remember that $\bar{\mathbf{s}}^{p+1}$ consists of all points $(x^1, \cdots, x^{p+1})$ satisfying

$$x^i \geqq 0, \qquad \sum_1^{p+1} x^i \leqq 1.$$

We have

$$\int_{\bar{\mathbf{s}}^{p+1}} d\eta = (-1)^p \int_{\bar{\mathbf{s}}^{p+1}} \frac{\partial A}{\partial x^{p+1}}\, dx^1 \cdots dx^{p+1}$$

$$= (-1)^p \int_{\{x^i \geqq 0,\ \sum_1^p x^i \leqq 1\}} dx^1 \cdots dx^p \left(\int_0^{(1-\sum_1^p x^i)} \frac{\partial A}{\partial x^{p+1}}\, dx^{p+1} \right)$$

$$= (-1)^p \int_{\{x^i \geqq 0,\ \sum_1^p x^i \leqq 1\}} \left[A(x^1, \cdots, x^p, 1 - \sum_1^p x^i) - A(x^1, \cdots, x^p, 0) \right] dx^1 \cdots dx^p.$$

We must next investigate $\partial\bar{\mathbf{s}}^{p+1}$. We write

$$\bar{\mathbf{s}}^{p+1} = (R_0, R_1, \cdots, R_{p+1}),$$

$$\left.\begin{aligned} R_0 &= 0 \\ R_1 &= (10 \cdots 0) \\ &\cdots\cdots\cdots \\ R_{p+1} &= (0 \cdots 01) \end{aligned}\right\} \quad \text{points in } \mathbf{E}^{p+1}.$$

We have

$$\partial\bar{\mathbf{s}}^{p+1} = (R_1, \cdots, R_{p+1}) + (-1)^{p+1}(R_0, R_1, \cdots, R_p)$$
$$+ \text{ other faces,}$$

where $\eta = 0$ on each of the other faces since some one of $x^1, \cdots, x^p$ is constant there. Thus

$$\int_{\partial \bar{s}^{p+1}} \eta = \int_{(R_1, \cdots, R_{p+1})} \eta + (-1)^{p+1} \int_{(R_0, R_1, \cdots, R_p)} \eta.$$

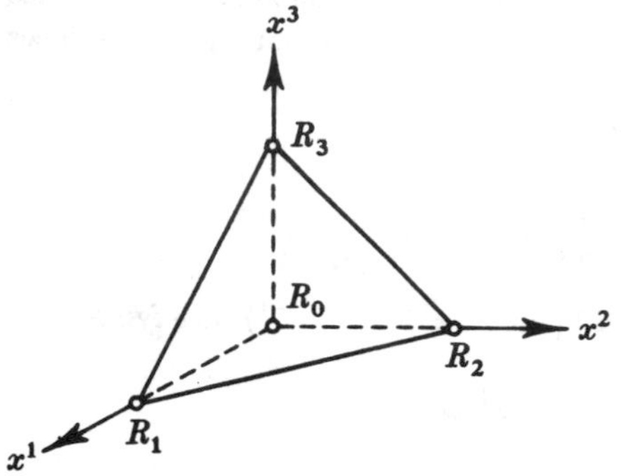

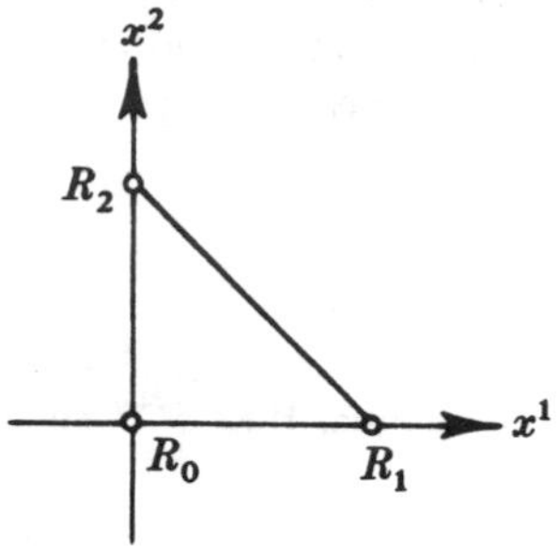

The face $(R_0, R_1, \cdots, R_p)$ is the standard $\bar{s}^p$. On it $x^{p+1} = 0$ and so

$$(-1)^{p+1} \int_{(R_0, \cdots, R_p)} \eta = (-1)^{p+1} \int_{\bar{s}^p} A(x^1, x^2, \cdots, x^p, 0)\, dx^1 \cdots dx^p$$

which is precisely the second term in the expression for $\int d\eta$ above. The first term is obtained by projecting downward in the x^{p+1} direction:

$$\int_{(R_1, \cdots, R_{p+1})} \eta = \int_{(R_1, \cdots, R_p, R_0)} A\left(x^1, \cdots, x^p, 1 - \sum_1^p x^i\right) dx^1 \cdots dx^p$$

$$= (-1)^p \int_{(R_0, R_1, \cdots, R_p)} A\left(x^1, \cdots, x^p, 1 - \sum_1^p x^i\right) dx^1 \cdots dx^p$$

$$= (-1)^p \int_{\bar{s}^p} A\left(x^1, \cdots, x^p, 1 - \sum_1^p x^i\right) dx^1 \cdots dx^p,$$

and this is the first term in the expression for $\int d\eta$. The proof is completed.

5.9. Periods and De Rham's Theorems

We consider an example. The manifold $\mathbf{M}$ consists of $\mathbf{E}^3$ with the origin removed,

$$\mathbf{M} = \mathbf{E}^3 - \{0\}.$$

Suppose ω is a one-form on $\mathbf{M}$ such that $d\omega = 0$. Then is ω exact? That is, is it the differential of a function on $\mathbf{M}$? The proof in Section 3.6 will not avail here because $\mathbf{M}$ cannot be shrunk to a point. Nonetheless, $\omega = df$, where

$$f(\mathbf{x}) = \int_{(1,0,0)}^{\mathbf{x}} \omega,$$

the integral taken along any path $\mathbf{c}$ which avoids 0. That this is independent of the path follows from Stokes' theorem. For if $\mathbf{c}'$ is another path in $\mathbf{M}$ from (1, 0, 0) to $\mathbf{x}$, then the chain $\mathbf{c} - \mathbf{c}'$ is the boundary of a piece of surface $\mathbf{\Sigma}$ (2-chain) in $\mathbf{M}$ and

$$\int_{\mathbf{c}} \omega - \int_{\mathbf{c}'} \omega = \int_{\mathbf{c}-\mathbf{c}'} \omega = \int_{\partial\Sigma} \omega = \int_{\Sigma} d\omega = 0.$$

Next suppose α is a two-form on $\mathbf{M}$ such that $d\alpha = 0$. We seek a one-form λ on $\mathbf{M}$ such that $\alpha = d\lambda$. By the converse to Poincaré's lemma in Section 3.6, such a form λ exists *locally*. But we are asking the *global* question: Is there such a form λ on *all* of $\mathbf{M}$? The answer to this one is no in general, we shall have explicit examples later. For if there were such a one-form λ with $d\lambda = \alpha$ we would have

$$\int_{\mathbf{S}^2} \alpha = \int_{\mathbf{S}^2} d\lambda = \int_{\partial\mathbf{S}^2} \lambda = 0$$

since the unit sphere $\mathbf{S}^2$ has no boundary. But there is no reason *a priori* for assuming that

$$\int_{\mathbf{S}^2} \alpha = 0.$$

The correct result is this. If α is a two-form on $\mathbf{M} = \mathbf{E}^3 - \{0\}$ with $d\alpha = 0$ and

$$\int_{\mathbf{S}^2} \alpha = 0,$$

then $\alpha = d\lambda$ for some one-form λ on $\mathbf{M}$.

This result is contained in De Rham's theorems which we shall formulate now without proofs.

We deal with a fixed manifold $\mathbf{M}$ about which we assume only some mild limitation on its size, for example we may suppose it can be imbedded in a sufficiently high dimensional Euclidean space.

A *closed form* is a differential form ω on $\mathbf{M}$ satisfying $d\omega = 0$.

An *exact form* is a differential form ω on $\mathbf{M}$ satisfying $\omega = d\eta$ for some form η on $\mathbf{M}$.

Each exact form is closed:

$$d\omega = d(d\eta) = 0.$$

Let ω be a closed p-form. To each p-cycle $\mathbf{z}$ on $\mathbf{M}$ corresponds a *period*† of ω,

$$\int_{\mathbf{z}} \omega.$$

If $\mathbf{z}$ happens to be a boundary $\mathbf{b} = \partial\mathbf{c}$, the period vanishes,

$$\int_{\mathbf{b}} \omega = \int_{\partial\mathbf{c}} \omega = \int_{\mathbf{c}} d\omega = \int_{\mathbf{c}} 0 = 0.$$

Because of this there is a relation between periods:

$$(\ddagger)\left\{\begin{array}{l}\textit{Whenever cycles } \mathbf{z}_1, \cdots \textit{ are related by} \\ \qquad\qquad \sum a_i\mathbf{z}_i = \textit{boundary}, \\ \textit{then} \\ \qquad\qquad \sum a_i \int_{\mathbf{z}_i} \omega = 0.\end{array}\right\}$$

De Rham's First Theorem. *A closed form is exact if and only if all of its periods vanish.*

De Rham's Second Theorem. *Suppose to each p-cycle* $\mathbf{z}$ *is assigned a number*, per($\mathbf{z}$), *subject to the consistency relations*

$$(\ddagger)\left\{\begin{array}{l}\textit{whenever} \\ \qquad\qquad \sum a_i\mathbf{z}_i = \textit{boundary}, \\ \textit{then} \\ \qquad\qquad \sum a_i \operatorname{per}(\mathbf{z}_i) = 0.\end{array}\right\}$$

Then there is a closed form ω *on* $\mathbf{M}$ *which has the assigned periods,*

$$\int_{\mathbf{z}} \omega = \operatorname{per}(\mathbf{z}) \qquad \text{for each } p\text{-cycle } \mathbf{z}.$$

On many spaces one is able to apply these results because there is a finite set of independent p-cycles which spans all p-cycles, up to boundaries. For example, on the n-sphere $\mathbf{S}^n$ it is known that each p-cycle is a boundary for $p > 0$, $p \neq n$, and that in dimension n there is a single n-cycle ($\mathbf{S}^n$ itself with outward normal for orientation) such that each n-cycle is a multiple of this one plus a boundary. These things are established by algebraic topology.

A complete analysis of De Rham's theorems reveals the following result, which has considerable attraction in itself.

Suppose we consider only chains $\mathbf{c} = \sum a_i\sigma_i$ which are sums of simplices

† The nomenclature derives from the periods of elliptic integrals and the corresponding differentials for algebraic functions.

with integer coefficients. Then we may talk of these as *integer-chains* and have *integer-cycles* and *integer-boundaries*. The *integer-periods*

$$\int_{\mathbf{z}} \omega$$

of a closed form ω are the periods taken over integer-cycles only.

Let ω and η be closed forms of degrees p and q respectively. Suppose that the integer-periods of ω and η are all integers. Then the same is true of $\omega \wedge \eta$.

5.10. Surfaces; Some Examples

It is shown in topology that each closed surface in $\mathbf{E}^3$ may be smoothly deformed into a sphere with h handles, or alternatively, a button with h holes. Let us consider the case $h = 2$ and orient this surface Σ with the outward drawn normal. The only significant two-cycle is Σ itself. By De Rham's First Theorem, a two-form α on this surface is an exact differential if and only if

$$\int_{\Sigma} \alpha = 0.$$

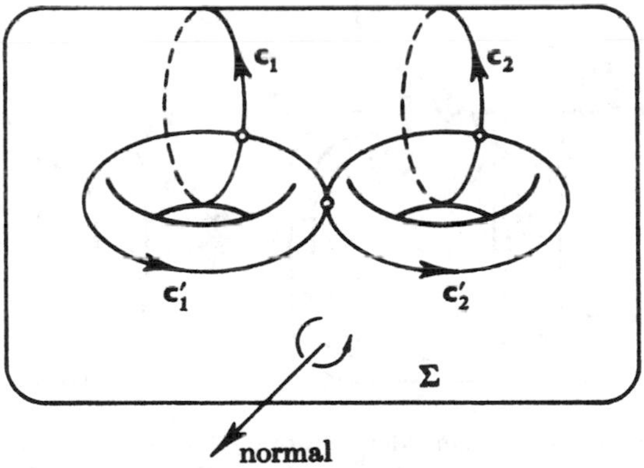

There are four significant one-cycles, $\mathbf{c}_1, \mathbf{c}_1', \mathbf{c}_2, \mathbf{c}_2'$. Here $\mathbf{c}_1$ and $\mathbf{c}_1'$ intersect once and cross, the same for $\mathbf{c}_2$ and $\mathbf{c}_2'$. But $\mathbf{c}_1'$ and $\mathbf{c}_2'$ intersect once without crossing. To see the geometric plausibility of the statement that each one-cycle $\mathbf{c}$ on Σ is a sum of multiples of the $\mathbf{c}_i$ and $\mathbf{c}_i'$ plus a boundary, one cuts the surface Σ along these basic cycles. Having done this, Σ may be smoothly deformed into a plane domain without holes.

De Rham's First Theorem now asserts that if ω is a closed one-form on Σ, then ω is an exact differential if and only if

$$\int_{\mathbf{c}_1} \omega = \int_{\mathbf{c}'_1} \omega = \int_{\mathbf{c}_2} \omega = \int_{\mathbf{c}'_2} \omega = 0.$$

Applied to dimension one, De Rham's Second Theorem asserts that if real numbers a_1, a_1', a_2, a_2' are given, there exists a closed one-form ω satisfying

$$\int_{\mathbf{c}_1} \omega = a_1, \qquad \int_{\mathbf{c}'_1} \omega = a_1', \qquad \int_{\mathbf{c}_2} \omega = a_2, \qquad \int_{\mathbf{c}'_2} \omega = a_2'.$$

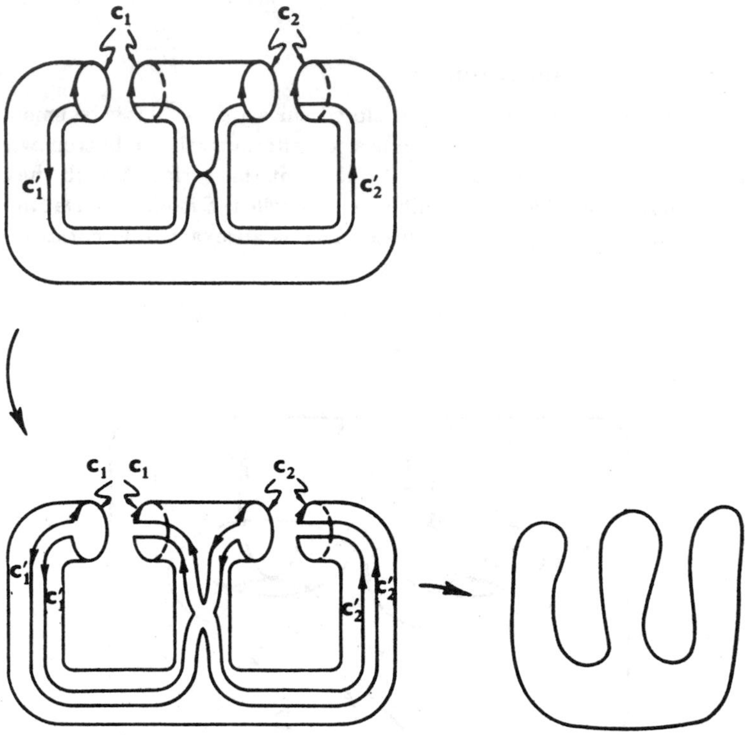

It is also interesting to consider non-orientable closed surfaces. These of course cannot be realized in $\mathbf{E}^3$. Perhaps the simplest is the projective plane $\mathbf{P}^2$. This is defined by pasting the edges of a rectangle together in the order indicated. The boundary relations are

$$\partial(\mathbf{P}^2) = 2\mathbf{c}' - 2\mathbf{c},$$
$$\partial\mathbf{c} = (P) - (Q)$$
$$\partial\mathbf{c}' = (P) - (Q).$$

This means first of all that there is no effective two-cycle, each two-form is exact. The only effective one-cycle is $\mathbf{c}' - \mathbf{c}$, and this actually bounds,

$\mathbf{c}' - \mathbf{c} = \frac{1}{2}\partial \mathbf{P}^2$. Thus each closed one-form is exact.

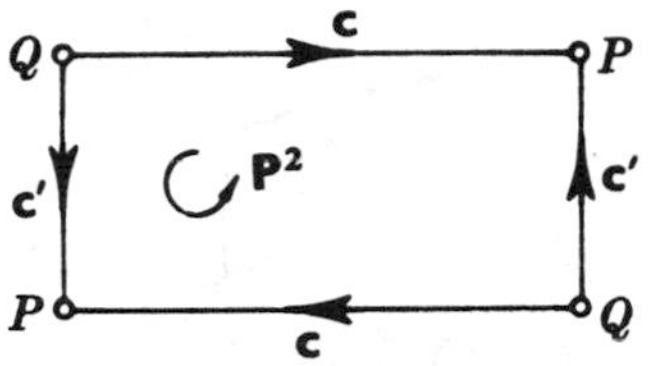

Another interesting example is the Klein bottle $\mathbf{K}^2$, again defined by pasting edges together. The boundary relations are

$$\partial \mathbf{K}^2 = -2\mathbf{c}$$

$$\partial \mathbf{c} = \partial \mathbf{c}' = 0.$$

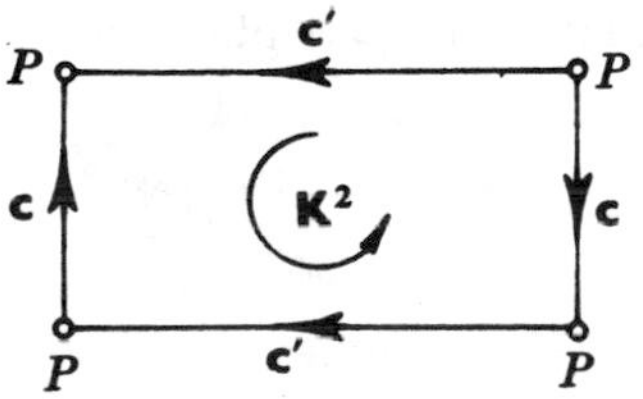

The one independent one-cycle is $\mathbf{c}'$.

5.11. Mappings of Chains

Suppose $\mathbf{M}$ and $\mathbf{N}$ are manifolds and f is a smooth mapping:

$$f\colon\ \mathbf{M} \longrightarrow \mathbf{N}.$$

Then to each p-chain $\mathbf{c}$ on $\mathbf{M}$ there corresponds in a natural way a p-chain $f_* \mathbf{c}$ on $\mathbf{N}$.

It suffices to explain this when $\mathbf{c}$ is a simplex σ^p. Such a simplex is represented by $(\mathbf{s}^p, \mathbf{U}, \phi)$ where $\mathbf{U}$ is a neighborhood of the Euclidean simplex $\mathbf{s}^p$ and $\phi\colon\ \mathbf{U} \longrightarrow \mathbf{M}$. We merely compose f and ϕ so that $f_* \mathbf{c}$ is represented by

$$(\mathbf{s}^p, \mathbf{U}, f \circ \phi).$$

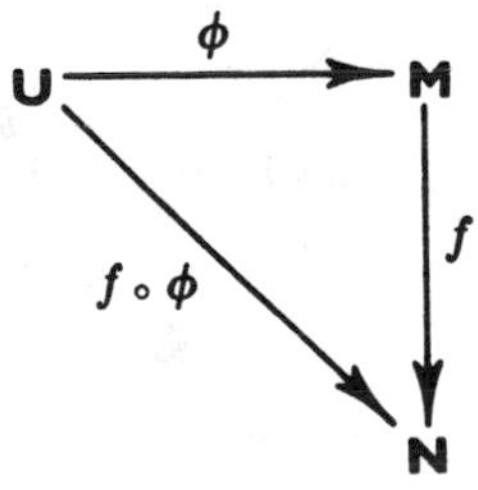

We illustrate the process for the case of a two-simplex

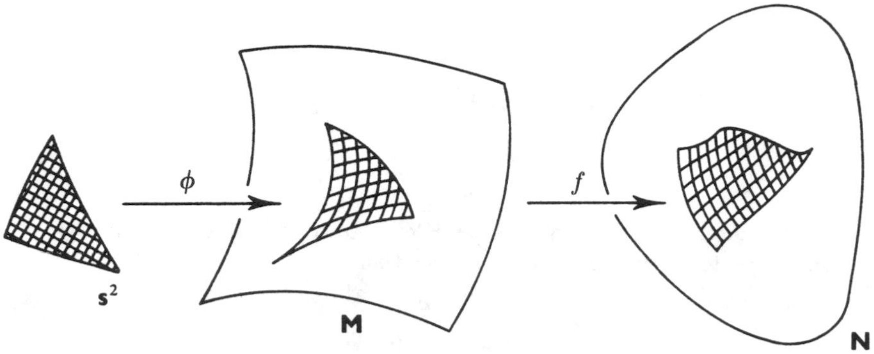

This induced map f_* takes the space of chains onto the space of chains:

$$\mathbf{M} \xrightarrow[f]{} \mathbf{N}$$

$$\mathbf{C}_p(\mathbf{M}) \xrightarrow[f_*]{} \mathbf{C}_p(\mathbf{N}).$$

We observe that if $\mathbf{c}$ is a p-chain in $\mathbf{M}$, then

$$f_*(\partial\mathbf{c}) = \partial(f_*\mathbf{c}),$$

which leads to the commutative diagram

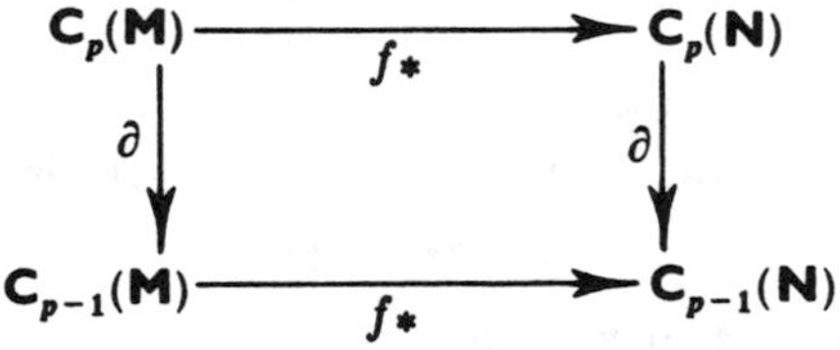

which is certainly analogous to the corresponding diagram in Section 5.4 for f^* and d. The validity of the result is established by looking at individual simplices.

We now see what happens with two mappings. Let

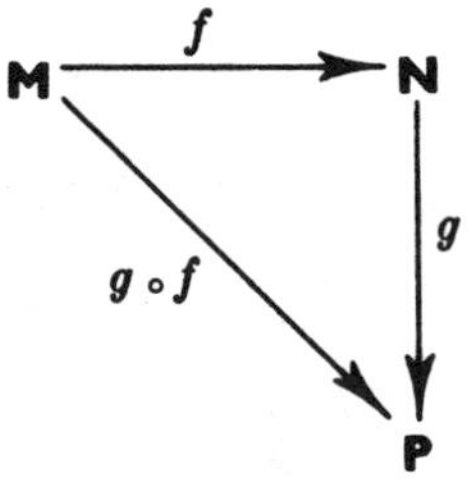

Then the assertion is

$$(g \circ f)_* = g_* \circ f_*$$

which again follows for a simplex almost directly from the definition of f_*.

Finally we consider this situation. Let

$$f\colon \mathbf{M} \longrightarrow \mathbf{N}.$$

Suppose that ω is a p-form on $\mathbf{N}$ and $\mathbf{c}$ is a p-chain on $\mathbf{M}$. Then $f^*\omega$ is a p-form on $\mathbf{M}$ and $f_*\mathbf{c}$ is a p-chain on $\mathbf{N}$. We have

$$\int_{\mathbf{c}} f^*\omega = \int_{f_*\mathbf{c}} \omega.$$

This important result also follows directly from the definition for a simplex and is obtained for a general chain by summation.

5.12. Problems

1. Show that the totality of unit tangent vectors to the sphere $\mathbf{S}^2$ is a three-manifold. Construct local coordinates.

2. Show that the set of all directed lines in $\mathbf{E}^2$ is a 2-manifold. Discuss orientation.

3. More generally, consider the set of all oriented r-dimensional planes in $\mathbf{E}^n$. Show that this is a manifold, compute its dimension, and discuss orientation.

4. Projective n-space $\mathbf{P}^n$ consists of all $(n + 1)$-tuples $(a_0, \cdots, a_n)$ of real numbers not all zero, where proportional $(n + 1)$-tuples are considered as representing the same point. Show that $\mathbf{P}^n$ is an n-manifold.

5. Complex projective n-space $\mathbf{CP}^n$ consists of all $(n + 1)$-tuples $(a_0, \cdots, a_n)$ of complex numbers not all zero, where two such n-tuples are considered the same if they differ by a (complex) proportionality factor. Show that $\mathbf{CP}^n$ is a manifold and determine its dimension.

6. Show that the manifolds of Examples 4 and 5 are closed (compact).

7. Let $\mathbf{M}$ be the manifold of Example 3. Show that the set $\mathbf{N}$ of all oriented r-planes in $\mathbf{E}^n$ which pass through the origin is a closed submanifold of $\mathbf{M}$.

8. Denote by $\mathbf{U}$ the open region

$$\mathbf{U} = \{x_1^2 + \cdots + x_n^2 > 1\}$$

in $\mathbf{E}^n$. Suppose ω is an r-form in $\mathbf{E}^n$ which vanishes identically on $\mathbf{U}$. Under what conditions does there exist an $(r - 1)$-form α on $\mathbf{E}^n$ which also vanishes identically on $\mathbf{U}$ and which satisfies $d\alpha = \omega$?

9. Show by direct calculation (i.e., without De Rham's theorems) that if ω is a two-form on $\mathbf{S}^2$ whose integral over $\mathbf{S}^2$ vanishes, then $\omega = d\alpha$ for a suitable one-form α on $\mathbf{S}^2$.

VI

Applications in Euclidean Space

6.1. Volumes in $\mathbf{E}^n$

We denote by

$$\omega = dx_1 \cdots dx_n$$

the element of volume in $\mathbf{E}^n$, an n-form, and set

$$V_n = \int_{r \leqq 1} \omega, \qquad r^2 = \sum x_i^2,$$

so that V_n is the volume of the unit ball. Next we denote by σ' the element of $(n-1)$-dimensional volume on the unit sphere $\mathbf{S}^{n-1} = \{\mathbf{x} \mid r = 1\}$, and set

$$A_{n-1} = \int_{\mathbf{S}^{n-1}} \sigma'.$$

Thus $A_1 = 2\pi$, $A_2 = 4\pi$, $V_1 = 2$, $V_2 = \pi$, $V_3 = \frac{4}{3}\pi$. It is clear that the volume of the sphere of radius r is $r^{n-1}A_{n-1}$, hence

$$V_n = \int_0^1 r^{n-1} A_{n-1}\, dr = \frac{1}{n} A_{n-1}.$$

One may evaluate V_n by integrating over slabs:

$$V_n = \int_{-1}^1 (1 - x^2)^{(n-1)/2} V_{n-1}\, dx$$

$$= V_{n-1} J_n$$

where

$$J_n = \int_{-1}^1 (1 - x^2)^{(n-1)/2}\, dx.$$

Integration by parts once leads to

$$J_n = \int_{-1}^1 x(2x)\left(\frac{n-1}{2}\right)(1 - x^2)^{(n-3)/2}\, dx = (n-1)(-J_n + J_{n-2}),$$

$$J_n = \frac{n-1}{n} J_{n-2}.$$

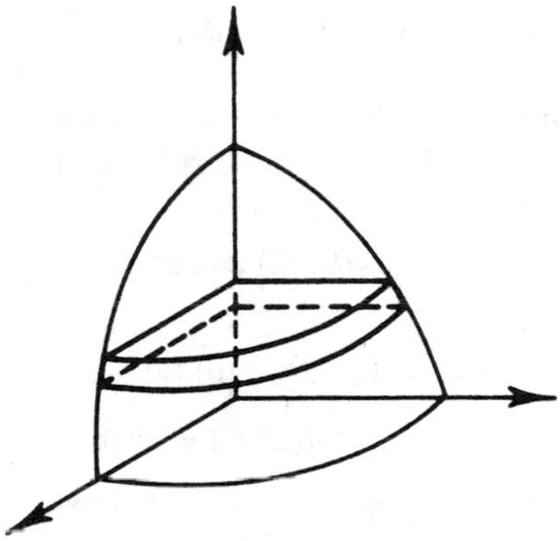

These recursion formulae lead to the standard result

$$V_n = \frac{\pi^{n/2}}{\Gamma\left(\frac{n}{2}+1\right)}.$$

Next we obtain an explicit formula for σ' in terms of the Euclidean coordinates $x_1, \cdots, x_n$. We begin with the form

$$r\,dr = \sum x_i\,dx_i,$$

a one-form in $\mathbf{E}^n$ which is invariant under rotations (orthogonal transformations) of $\mathbf{E}^n$. Consequently

$$*\,r\,dr = \sum (-1)^{i-1} x_i\,dx_1 \cdots \widehat{dx_i} \cdots dx_n$$

(the "hat" denotes a missing factor) is an $(n-1)$-form in $\mathbf{E}^n$ which is invariant under rotations. It follows that on $\mathbf{S}^{n-1}$,

$$\sigma' = c * r\,dr,$$

where c is a constant.

Next we note that

$$d(*\,r\,dr) = \sum (-1)^{i-1} dx_i\,dx_1 \cdots \widehat{dx_i} \cdots dx_n = n\omega,$$

hence

$$\begin{aligned} A_{n-1} &= \int_{\mathbf{S}^{n-1}} \sigma' = c\int_{\mathbf{S}^{n-1}} *\,r\,dr = c\int_{r\leqq 1} d(*\,r\,dr) \\ &= c\int_{r\leqq 1} n\omega = cnV_n = cA_{n-1}, \end{aligned}$$

$c = 1$, $\sigma' = *\,r\,dr$ on $\mathbf{S}^{n-1}$.

Summarizing, we set

$$\sigma = * r\,dr = \sum_{i=1}^{n} (-1)^{i-1} x_i\, dx_1 \cdots \widehat{dx_i} \cdots dx_n,$$

defining an $(n-1)$-form σ in $\mathbf{E}^n$. Then $d\sigma = n\omega$, and if σ is restricted to $\mathbf{S}^{n-1}$, the result is the $(n-1)$-dimensional volume form σ' of $\mathbf{S}^{n-1}$.

Next we consider the natural projection

$$\pi\colon\ \mathbf{E}^n - \{0\} \longrightarrow \mathbf{S}^{n-1}$$

defined by $\pi(\mathbf{x}) = \mathbf{x}/|\mathbf{x}|$.

We seek $\pi^* \sigma'$, an $(n-1)$-form on $\mathbf{E}^n - \{0\}$ satisfying

$$d(\pi^* \sigma') = 0 \qquad \text{since} \qquad d(\pi^* \sigma') = \pi^* (d\sigma') = \pi^* (0) = 0.$$

($d\sigma'$ is an n-form on $\mathbf{S}^{n-1}$, hence 0.) We shall prove

$$\pi^* \sigma' = \frac{\sigma}{r^n}.$$

We could prove this by directly substituting

$$y_i = x_i/r \qquad \text{in} \qquad \sigma' = \sum (-1)^{i-1} y_i\, dy_1 \cdots \widehat{dy_i} \cdots dy_n,$$

but we prefer to proceed indirectly by exploiting the symmetries present. We set

$$\tau = \frac{\sigma}{r^n}.$$

Then

$$d\tau = \frac{1}{r^n}\, d\sigma - \frac{n}{r^{n+2}} (r\,dr)\sigma = \frac{n\omega}{r^n} - \frac{nr^2\omega}{r^{n+2}} = 0.$$

Now we observe that $*(\pi^* \sigma')$ is a one-form in $\mathbf{E}^n - \{0\}$ which is invariant under rotations, hence dependent on r alone. We may write

$$*(\pi^* \sigma') = \frac{f(r)}{r^n} (r\,dr).$$

From this we have

$$\pi^* \sigma' = \frac{f(r)}{r^n}\, \sigma = f(r)\tau,$$

$$0 = d(\pi^* \sigma') = \frac{df}{dr}\, \tau, \qquad \frac{df}{dr} = 0, \qquad f = c,$$

a constant, $\pi^* \sigma' = c\tau$. To evaluate c, we simply note that on $\mathbf{S}^{n-1}$, both $\pi^* \sigma'$ and τ collapse to σ, hence $c = 1$,

$$\pi^* \sigma' = \tau.$$

6.2. Winding Numbers, Degree of a Mapping

A basic result of topology (Seifert und Threlfall [20], p. 283) asserts that if $\mathbf{M}$ and $\mathbf{N}$ are closed oriented n-manifolds and $f\colon \mathbf{M} \longrightarrow \mathbf{N}$, then the chain $f_* \mathbf{M}$ is an integral multiple of $\mathbf{N}$ plus a boundary. This integer multiplier is called the *degree* of f and written $\deg f$.

Now suppose that Σ is a closed oriented $(n-1)$-manifold in $\mathbf{E}^n - \{0\}$. Then by the Jordan–Brouwer theorem of topology, Σ decomposes $\mathbf{E}^n$ into exactly two regions. We assume Σ is oriented by the outward normal. The projection mapping π of Section 6.1 sends Σ into $\mathbf{S}^{n-1}$. It is true that $\deg \pi = 0$ or 1; our point is that this can be determined by an integral. Let $\delta = \deg \pi$.

Then

$$\int_\Sigma \tau = \int_\Sigma \pi^* \sigma' = \int_{\pi(\Sigma)} \sigma' = \delta \int_{\mathbf{S}^{n-1}} \sigma' = \delta A_{n-1},$$

hence

$$\delta = \frac{1}{A_{n-1}} \int_\Sigma \tau.$$

More generally, let $\mathbf{M}^{n-1}$ be a closed oriented manifold,

$$f\colon \quad \mathbf{M}^{n-1} \longrightarrow \mathbf{E}^n - \{0\}.$$

Essentially we are thinking of $f(\mathbf{M}^{n-1})$ as a hypersurface in $\mathbf{E}^n - \{0\}$ which may intersect itself We look on this hypersurface as winding around the origin and we want to count how many times it encircles. This winding number is given by the *Kronecker integral*

$$\frac{1}{A_{n-1}} \int_{\mathbf{M}} f^* \tau.$$

We may justify this as follows. Set $g = \pi \circ f\colon \ \mathbf{M}^{n-1} \longrightarrow \mathbf{S}^{n-1}$. What we are after is $\deg g$. Now

$$g_*(\mathbf{M}) = (\deg g)\mathbf{S}^{n-1} + (\text{boundary}),$$

hence

$$\begin{aligned} \int_{\mathbf{M}} g^* \sigma' &= \int_{g_* \mathbf{M}} \sigma' \\ &= (\deg g) \int_{\mathbf{S}^{n-1}} \sigma' = A_{n-1} \deg g, \end{aligned}$$

$$\deg g = \frac{1}{A_{n-1}} \int_{\mathbf{M}} g^* \sigma'.$$

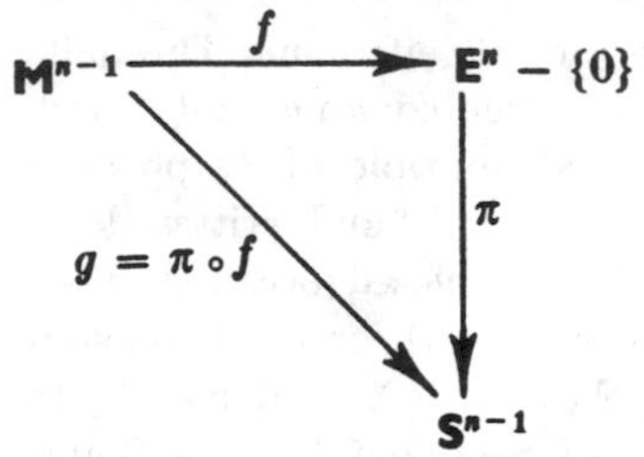

But $g^* \sigma' = (f^* \circ \pi^*)\sigma' = f^* \tau$, so we have

$$\deg g = \frac{1}{A_{n-1}} \int_{\mathbf{M}} f^* \tau.$$

The most general situation is this:

$$f\colon \quad \mathbf{M}^n \longrightarrow \mathbf{N}^n.$$

Let β be the volume form on $\mathbf{N}$ taken so that $\int_{\mathbf{N}} \beta = 1$. Then

$$\deg f = \int_{\mathbf{M}} f^* \beta.$$

For $f_* \mathbf{M} = (\deg f)\mathbf{N} + (\text{boundary})$, hence

$$\int_{\mathbf{M}} f^* \beta = \int_{f_* \mathbf{M}} \beta = (\deg f) \int_{\mathbf{N}} \beta = \deg f.$$

One interesting example: let $\mathbf{T}^n$ be the n-torus, $f\colon \mathbf{S}^n \longrightarrow \mathbf{T}^n$ where $n \geqq 2$. Then $\deg f = 0$.

Because the integrals involved are integer-valued, they remain constant when the mapping in question is subject to a deformation. Precisely, let $f_t\colon \mathbf{M} \longrightarrow \mathbf{N}$ be a one-parameter family of maps. Then

$$\deg f_t = \int_{\mathbf{M}} f_t{}^* \beta$$

is a smooth function of t, always an integer, hence constant. It follows that $\deg f_0 = \deg f_1$.

One other remark. Suppose we have

$$f\colon \quad \mathbf{M} \longrightarrow \mathbf{N}, \qquad g\colon \quad \mathbf{N} \longrightarrow \mathbf{P} \qquad \text{so that} \qquad h = g \circ f\colon \quad \mathbf{M} \longrightarrow \mathbf{P}.$$

Then

$$\deg h = (\deg f) \cdot (\deg g).$$

6.3. The Hopf Invariant

For each sphere $\mathbf{S}^n$, let σ_n denote the element of area, normalized so that

$$\int_{\mathbf{S}^n} \sigma_n = 1.$$

Consider first a map $f\colon \mathbf{S}^3 \longrightarrow \mathbf{S}^2$. Then $f^*\sigma_2$ is a 2-form on $\mathbf{S}^3$. Also $d(f^*\sigma_2) = f^*(d\sigma_2) = 0$. Since $\mathbf{S}^3$ has no nontrivial 2-dimensional cycles, we deduce that

$$f^*\sigma_2 = d\alpha_1$$

where the one-form α_1 on $\mathbf{S}^3$ is unique up to the differential of a function. The 3-form $\alpha_1 \wedge f^*\sigma_2$ has an integral

$$\int_{\mathbf{S}^3} \alpha_1 \wedge f^*\sigma_2,$$

which has the remarkable property of being an integer, called the *Hopf invariant* of f. It is invariant under deformation of f. More generally, let

$$f\colon \mathbf{S}^{2n-1} \longrightarrow \mathbf{S}^n.$$

Then $f^*\sigma_n = d\alpha_{n-1}$, and the Hopf invariant of f is

$$\int_{\mathbf{S}^{2n-1}} \alpha_{n-1} \wedge f^*\sigma_n.$$

We may represent $\mathbf{S}^3$ by pairs of complex numbers

$$(z, w), \qquad |z|^2 + |w|^2 = 1.$$

The mapping $(z, w) \longrightarrow z/w$ provides a mapping of $\mathbf{S}^3$ into the closed complex plane, i.e., the Riemann sphere $\mathbf{S}^2$. This map has Hopf invariant $+1$, hence it is *essential* in the sense that it cannot be deformed to a trivial map, everything going to a single point.

6.4. Linking Numbers, The Gauss Integral, Ampère's Law

Let $\mathbf{M}^r$, $\mathbf{N}^s$ be oriented closed manifolds in $\mathbf{E}^n$, where $r + s = n - 1$, and suppose these have no common point. (Best example: two disjoint closed curves in $\mathbf{E}^3$.) We want to count how many times they link. To do this, we form the product space $\mathbf{M} \times \mathbf{N}$ which is an oriented manifold of dimension $r + s = n - 1$. We consider the map $f\colon \mathbf{M} \times \mathbf{N} \longrightarrow \mathbf{E}^n - \{0\}$ defined by

$$f(\mathbf{x}, \mathbf{y}) = \mathbf{y} - \mathbf{x}.$$

Now we set

$$\operatorname{link}(\mathbf{M}, \mathbf{N}) = \deg f.$$

Thus if τ is the n-form in $\mathbf{E}^n - \{0\}$ we considered above,

$$\operatorname{link}(\mathbf{M}, \mathbf{N}) = \frac{1}{A_{n-1}} \int_{\mathbf{M} \times \mathbf{N}} f^* \tau = \frac{1}{A_{n-1}} \int_{\mathbf{M}} \int_{\mathbf{N}} f^* \tau.$$

We shall work this out in $\mathbf{E}^3$ for a pair of closed curves $\mathbf{M}$, $\mathbf{N}$:

$$\tau = \frac{1}{|\mathbf{z}|^3} \sum z_i \, dz_j \, dz_k = \frac{1}{|\mathbf{z}|^3} \frac{(\mathbf{z} \times d\mathbf{z}) \cdot d\mathbf{z}}{2}.$$

We let $\mathbf{x}$, $\mathbf{y}$ be the moving points on $\mathbf{M}$, $\mathbf{N}$, respectively. Then

$$\mathbf{z} = f(\mathbf{x}, \mathbf{y}) = \mathbf{y} - \mathbf{x}$$

so that

$$\begin{aligned} f^* \tau &= \frac{1}{2|\mathbf{y} - \mathbf{x}|^3} [(\mathbf{y} - \mathbf{x}) \times (d\mathbf{y} - d\mathbf{x})] \cdot (d\mathbf{y} - d\mathbf{x}) \\ &= \frac{1}{2|\mathbf{y} - \mathbf{x}|^3} \{[-(\mathbf{y} - \mathbf{x}) \times d\mathbf{y}] \cdot d\mathbf{x} - [(\mathbf{y} - \mathbf{x}) \times d\mathbf{x}] \cdot d\mathbf{y}\} \\ &= -\frac{1}{|\mathbf{y} - \mathbf{x}|^3} [(\mathbf{y} - \mathbf{x}) \times d\mathbf{y}] \cdot d\mathbf{x}, \end{aligned}$$

$$\operatorname{link}(\mathbf{M}, \mathbf{N}) = \frac{-1}{4\pi} \int_{\mathbf{x} \in \mathbf{M}} d\mathbf{x} \cdot \int_{\mathbf{y} \in \mathbf{N}} \frac{(\mathbf{y} - \mathbf{x}) \times d\mathbf{y}}{|\mathbf{y} - \mathbf{x}|^3}.$$

(In this computation $d\mathbf{y} \times d\mathbf{y} = 0$, etc., since $d\mathbf{y}$ involves only one variable.)

Imagine a steady unit electric current flowing around the closed loop $\mathbf{N}$. By Ampère's law, the magnetic field at a point $\mathbf{x}$ due to the current in a segment $d\mathbf{y}$ is

$$-\frac{1}{4\pi} \frac{(\mathbf{y} - \mathbf{x}) \times d\mathbf{y}}{|\mathbf{y} - \mathbf{x}|^3},$$

hence the total magnetic field at $\mathbf{x}$ is

$$\mathbf{F}(\mathbf{x}) = -\frac{1}{4\pi} \int_{\mathbf{N}} \frac{(\mathbf{y} - \mathbf{x}) \times d\mathbf{y}}{|\mathbf{y} - \mathbf{x}|^3}.$$

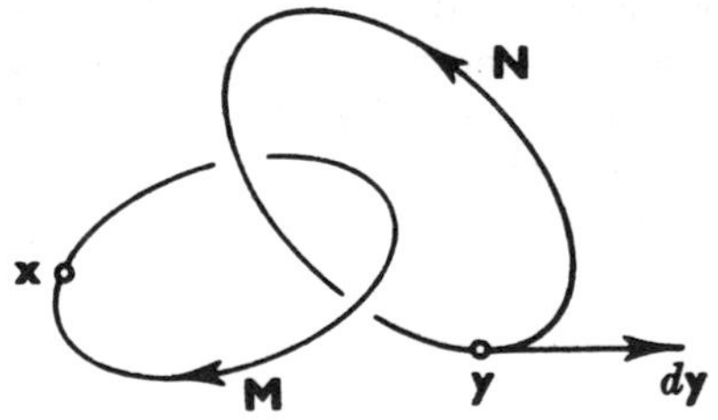

It follows that link $(\mathbf{M}, \mathbf{N}) = \int_{\mathbf{M}} \mathbf{F}(\mathbf{x}) \cdot d\mathbf{x}$ is precisely the work done by this field on a unit magnetic pole which makes one circuit of $\mathbf{M}$.

In the next example, link $(\mathbf{M}, \mathbf{N}) = 0$, which seems surprising since the curves cannot really be separated.

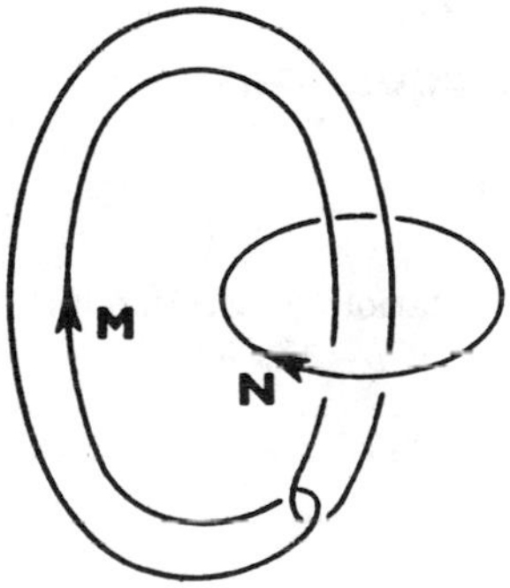

VII

Applications to Differential Equations

7.1. Potential Theory

We summarize the notation of Section 6.1. Space: $\mathbf{E}^n$. Coordinates: $x_1, x_2, \cdots, x_n$.

$$r^2 = x_1^2 + \cdots + x_n^2,$$

$$r\,dr = \sum_1^n x_i\,dx_i,$$

$$\sigma = *(r\,dr) = \sum_1^n x_i *dx_i = \sum_1^n (-1)^{i-1} x_i\,dx_1 \cdots \widehat{dx_i} \cdots dx_n.$$

$$\tau = \frac{\sigma}{r^n},$$

$$\omega = dx_1 \cdots dx_n = \text{volume element of } \mathbf{E}^n.$$

$$d\sigma = n\omega, \qquad d\tau = 0.$$

$$V_n = \int_{r \leqq 1} \omega = \frac{\pi^{n/2}}{\Gamma((n/2)+1)}, \qquad A_{n-1} = \int_{r=1} \sigma = nV_n.$$

Let u be a smooth function on a domain in $\mathbf{E}^n$. Then

$$du = \sum \frac{\partial u}{\partial x^i}\,dx^i,$$

$$*du = \sum (-1)^{i-1} \frac{\partial u}{\partial x^i}\,dx^1 \cdots \widehat{dx^i} \cdots dx^n,$$

$$d*du = \left(\sum \frac{\partial^2 u}{\partial x^{i2}}\right)\omega = (\Delta u)\omega,$$

defining the *Laplacian*

$$\Delta u = \sum \frac{\partial^2 u}{\partial x^{i2}}.$$

(See Section 4.4 for details when $n = 3$.)

If u and v are functions on the finite domain $\mathbf{R}$, the *Dirichlet* (*bilinear*) *integral* is

$$D[u, v] = \int_{\mathbf{R}} du \wedge *dv = \int_{\mathbf{R}} dv \wedge *du = \int_{\mathbf{R}} \sum_i \left(\frac{\partial u}{\partial x^i}\right)\left(\frac{\partial v}{\partial x^i}\right)\omega.$$

Next we have by Stokes' theorem

$$\int_{\partial\mathbf{R}} u *dv = \int_{\mathbf{R}} d(u *dv).$$

But

$$\begin{aligned} d(u *dv) &= du \wedge *dv + u\, d *dv \\ &= du \wedge *dv + u\, \Delta v\, \omega, \end{aligned}$$

hence we have

Green's Formula

$$\int_{\partial\mathbf{R}} u *dv = D[u, v] + \int_{\mathbf{R}} u\, \Delta v\, \omega.$$

By reversing u and v and subtracting the results, we obtain

Green's Symmetrical Formula

$$\int_{\partial\mathbf{R}} (u *dv - v *du) = \int_{\mathbf{R}} (u\, \Delta v - v\, \Delta u)\omega.$$

[One usually writes $*du = (\partial u/\partial \nu)\lambda$ where λ is the $(n-1)$-dimensional volume element on $\partial\mathbf{R}$ and $\partial u/d\nu$ is the *normal derivative*.]

In case v is harmonic in the region $\mathbf{R}$, $\Delta v = 0$, and we have

$$\int_{\partial\mathbf{R}} (u *dv - v *du) + \int_{\mathbf{R}} v\, \Delta u\, \omega = 0.$$

By specializing further we have this result:
Let u and v be harmonic functions in a region $\mathbf{R}$. Then

$$\int_{\partial\mathbf{R}} u *dv = \int_{\partial\mathbf{R}} v *du.$$

We derive further consequences by setting

$$\begin{aligned} v &= \frac{1}{r^{n-2}}, \\ dv &= \frac{-(n-2)}{r^n}(r\, dr), \\ *dv &= -(n-2)\tau, \\ d *dv &= 0, \qquad \Delta v = 0. \end{aligned}$$

The function v is defined on $\mathbf{E}^n - \{0\}$. We suppose the region $\mathbf{R}$ contains 0 and we apply the formula above to the punctured region

$$\mathbf{R} - \{r \leqq \varepsilon\}$$

with ε a small positive constant. We suppose u is harmonic on all of $\mathbf{R}$. Since

$$\partial[\mathbf{R} - \{r \leqq \varepsilon\}] = \partial\mathbf{R} - \{r = \varepsilon\},$$

we have

$$\int_{\partial\mathbf{R}} u *dv - \int_{r=\varepsilon} u *dv = \int_{\partial\mathbf{R}} v *du - \int_{r=\varepsilon} v *du,$$

$$-(n-2)\int_{\partial\mathbf{R}} u\tau + (n-2)\int_{r=\varepsilon} u\tau = \int_{\partial\mathbf{R}} \frac{1}{r^{n-2}} *du - \int_{r=\varepsilon} \frac{1}{r^{n-2}} *du.$$

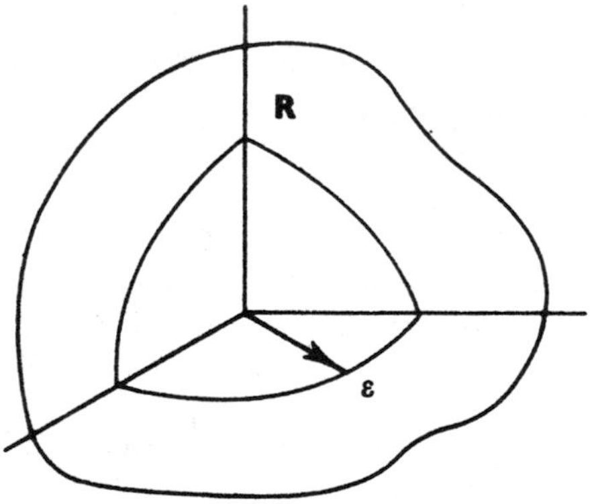

We evaluate the individual terms:

$$\int_{r=\varepsilon} u\tau = \frac{1}{\varepsilon^n}\int_{r=\varepsilon} u\sigma = \frac{1}{\varepsilon^n}\int_{r\leqq\varepsilon} d(u\sigma) = \frac{1}{\varepsilon^n}\int_{r\leqq\varepsilon} (du \wedge \sigma + u\cdot n\omega).$$

Now for $|\mathbf{x}| \leqq \varepsilon$, $u(\mathbf{x}) = u(0) + O(\varepsilon)$, hence

$$\int_{r\leqq\varepsilon} u\omega = u(0)V_n + O(\varepsilon)\varepsilon^n.$$

Similarly

$$\int_{r\leqq\varepsilon} du \wedge \sigma = \int_{r\leqq\varepsilon}\left(\sum \frac{\partial u}{\partial x^i}x^i\right)\omega = \int_{r\leqq\varepsilon} O(\varepsilon)\omega = O(\varepsilon)\varepsilon^n,$$

hence

$$\int_{r=\varepsilon} u\tau = nV_n u(0) + O(\varepsilon) = A_{n-1}u(0) + O(\varepsilon).$$

$$\int_{r=\varepsilon} \frac{1}{r^{n-2}} *du = \frac{1}{\varepsilon^{n-2}} \int_{r=\varepsilon} *du = \frac{1}{\varepsilon^{n-2}} \int_{r\leqq\varepsilon} d(*du)$$

$$= \frac{1}{\varepsilon^{n-2}} \int_{r\leqq\varepsilon} (\Delta u)\omega = 0.$$

We substitute these results and let $\varepsilon \longrightarrow 0$ to get

$$u(0) = \frac{1}{A_{n-1}} \int_{\partial\mathbf{R}} u\tau + \frac{1}{(n-2)A_{n-1}} \int_{\partial\mathbf{R}} \frac{*du}{r^{n-2}},$$

which gives the value of a harmonic function at a point in terms of the boundary values of it and its normal derivative.

Special case. Let $\mathbf{R}$ be the spherical region of radius a centered at 0, $\mathbf{R} = \{r \leq a\}$. In this case the second term on the right-hand side vanishes for the same reason that the corresponding integral taken over $\{r = \varepsilon\}$ vanishes. On $\partial\mathbf{R} = \{r = a\}$ we have

$$\tau = \frac{\sigma}{a^n} = \frac{1}{a^{n-1}} \mu$$

where

$$\mu = \mu_{\mathbf{x}} = \frac{1}{a} \sum (-1)^{i-1} x_i\, dx_1 \cdots \widehat{dx_i} \cdots dx_n$$

is the element of $(n-1)$-dimensional volume on $\{r = a\}$. (For $a = 1$ this reduces to σ. Since there are n x-terms in the numerator and $a = |\mathbf{x}|$ is in the denominator, it is homogeneous of the right degree, $n - 1$.) We have the

Gauss Mean Value Theorem

$$u(0) = \frac{1}{a^{n-1}A_{n-1}} \int_{r=a} u\mu = \frac{\int_{r=a} u\mu}{\int_{r=a} \mu},$$

which tells us that the mean value of u over the sphere of radius a is the value of u at the center.

Two important properties of harmonic functions follow from this result.

Maximum (Minimum) Principle. *Let u be harmonic on the finite region $\mathbf{R}$. Then u never assumes its maximum (minimum) value at an interior point of $\mathbf{R}$ unless u is constant.*

For suppose u assumes its maximum at an interior point x_0 of $\mathbf{R}$ which, after translation of coordinates, may be taken to be 0. Let Σ be any $(n-1)$-sphere of radius a centered at 0 with a so small that $\{r \leqq a\}$ is in $\mathbf{R}$. Since $u(\mathbf{x}) \leqq u(0)$ we have

$$u(0) = \int_{\Sigma} u\mu \Big/ \int_{\Sigma} \mu \leqq \int_{\Sigma} u(0)\mu \Big/ \int_{\Sigma} \mu = u(0)$$

so we must have $u(\mathbf{x}) = u(0)$ for all $\mathbf{x}$ in Σ. Hence u is constant on the largest spherical neighborhood of $\mathbf{x}_0$ we can draw in $\mathbf{R}$. Evidently this means u is constant in all of $\mathbf{R}$ since we can reach any other interior point by a sequence of such overlapping spheres. The result for minima follows the same way.

Uniqueness Principle for the Boundary Value Problem. *Let u and v be harmonic on a finite domain $\mathbf{R}$ and coincide on $\partial\mathbf{R}$. Then $u = v$ on $\mathbf{R}$.*

For $u - v$ vanishes on $\partial\mathbf{R}$ and is harmonic. By the Maximum Principle, $u - v \leqq 0$ on $\mathbf{R}$, $u \leqq v$. Similarly $v \leqq u$, hence $u = v$.

The function $(1/r^{n-2})$ is ideally suited to the sphere. On other domains it is inconvenient because of the term involving the normal derivative in the expression above for $u(0)$. Hence we introduce the Green's function.

Let $\mathbf{R}$ be a finite domain. A function $v(\mathbf{x}, \mathbf{y})$ defined for $\mathbf{x}$ and $\mathbf{y}$ distinct points of $\mathbf{R}$ is called the *Green's function* of $\mathbf{R}$ if

(i) For each fixed $\mathbf{y}$ in $\mathbf{R}$, $v(\mathbf{x}, \mathbf{y})$ is a harmonic function of $\mathbf{x}$ for $\mathbf{x}$ in $\mathbf{R} - \{\mathbf{y}\}$.

(ii) For each fixed $\mathbf{y}$ in $\mathbf{R}$, $v(\mathbf{x}, \mathbf{y}) = 0$ for $\mathbf{x}$ in $\partial\mathbf{R}$.

(iii) For each fixed $\mathbf{y}$ in $\mathbf{R}$,

$$v(\mathbf{x}, \mathbf{y}) - \frac{1}{|\mathbf{x} - \mathbf{y}|^{n-2}}$$

is a smooth harmonic function on all of $\mathbf{R}$.

Using the same method as above one proves the following:

If u is any harmonic function on a finite domain $\mathbf{R}$ and $v = v(\mathbf{x}, \mathbf{y})$ is the Green's function for $\mathbf{R}$, then

$$u(\mathbf{y}) = \frac{-1}{(n-2)A_{n-1}} \int_{\partial\mathbf{R}} u(\mathbf{x}) *d_{\mathbf{x}}v(\mathbf{x}, \mathbf{y}).$$

In case $\mathbf{R}$ is the spherical domain $r \leqq a$ centered at 0, the Green's function is

$$v(\mathbf{x}, \mathbf{y}) = \frac{1}{|\mathbf{x} - \mathbf{y}|^{n-2}} - \frac{a^{n-2}}{|\mathbf{y}|^{n-2}\left|\mathbf{x} - \dfrac{a^2}{|\mathbf{y}|^2}\mathbf{y}\right|^{n-2}}.$$

We note that

$$\mathbf{y}' = \frac{a^2}{|\mathbf{y}|^2}\,\mathbf{y}$$

is the inverse of $\mathbf{y}$ with respect to the sphere $\mathbf{R}$ (reciprocal radii). For $\mathbf{x}$ on $\partial\mathbf{R}$, $|\mathbf{x}| = a$ and we have

$$|\mathbf{x} - \mathbf{y}'|^2 = (\mathbf{x} - \mathbf{y}')\cdot(\mathbf{x} - \mathbf{y}') = a^2 - 2\frac{a^2}{|\mathbf{y}|^2}\,\mathbf{x}\cdot\mathbf{y} + \frac{a^4}{|\mathbf{y}|^2}$$

$$= \frac{a^2}{|\mathbf{y}|^2}\,(|\mathbf{y}|^2 - 2\mathbf{x}\cdot\mathbf{y} + |\mathbf{x}|^2) = \frac{a^2}{|\mathbf{y}|^2}\,|\mathbf{x} - \mathbf{y}|^2,$$

$$|\mathbf{x} - \mathbf{y}'| = \frac{a}{|\mathbf{y}|}\,|\mathbf{x} - \mathbf{y}|.$$

This explains why v vanishes for $\mathbf{x}$ on $\partial\mathbf{R}$.

Next

$$*d_{\mathbf{x}}v = -(n-2)\left[\tau(\mathbf{x} - \mathbf{y}) - \frac{a^{n-2}}{|\mathbf{y}|^{n-2}}\,(\mathbf{x} - \mathbf{y}')\right]$$

where

$$\tau(\mathbf{x} - \mathbf{y}) = \frac{1}{|\mathbf{x} - \mathbf{y}|^n}\sum (x_i - y_i)\,*dx_i\,,$$

$$\tau(\mathbf{x} - \mathbf{y}') = \frac{1}{|\mathbf{x} - \mathbf{y}'|^n}\sum (x_i - y_i')\,*dx_i\,.$$

We only need dv for $\mathbf{x}$ on $\partial\mathbf{R}$. Recalling that $\mathbf{y}' = (a^2/|\mathbf{y}|^2)\mathbf{y}$ for $|\mathbf{x}| = a$ we have

$$*d_{\mathbf{x}}v\Big|_{|\mathbf{x}|=a} = \frac{-(n-2)}{|\mathbf{x} - \mathbf{y}|^n}\sum\left[(x_i - y_i) - \frac{|\mathbf{y}|^2}{a^2}\left(x_i - \frac{a^2}{|\mathbf{y}|^2}\,y_i\right)\right]*dx_i$$

$$= \frac{-(n-2)}{|\mathbf{x} - \mathbf{y}|^n}\,\frac{a^2 - |\mathbf{y}|^2}{a^2}\,\sigma = \frac{-(n-2)}{|\mathbf{x} - \mathbf{y}|^n}\,\frac{a^2 - |\mathbf{y}|^2}{a}\,\mu_{\mathbf{x}},$$

so that the representation of u in terms of its boundary values specializes to

$$u(\mathbf{y}) = \frac{a^2 - |\mathbf{y}|^2}{aA_{n-1}}\int_{|\mathbf{x}|=a}\frac{u(\mathbf{x})}{|\mathbf{x} - \mathbf{y}|^n}\,\mu_{\mathbf{x}}.$$

This is the Poisson Integral Formula which provides an explicit solution formula for the Dirichlet problem on the sphere—find a harmonic function with prescribed boundary values.

Returning to a general finite domain, we mention the important symmetry property of the Green's function:

$$v(\mathbf{x}, \mathbf{y}) = v(\mathbf{y}, \mathbf{x}).$$

(That this is the case for the sphere is not apparent from the unsymmetrical formula above. But for the denominator of the second term we have

$$|\mathbf{y}|^2|\mathbf{x}-\mathbf{y}'|^2 = |\mathbf{y}|^2(\mathbf{x}-\mathbf{y}')\cdot(\mathbf{x}-\mathbf{y}') = |\mathbf{y}|^2\left(|\mathbf{x}|^2 - 2\frac{a^2}{|\mathbf{y}|^2}(\mathbf{x}\cdot\mathbf{y}) + \frac{a^4}{|\mathbf{y}|^2}\right)$$

$$= |\mathbf{x}|^2|\mathbf{y}|^2 - 2a^2(\mathbf{x}\cdot\mathbf{y}) + a^4,$$

which turns out to be symmetrical after all.)

One of the many important consequences of the Poisson Integral Formula is the

LIOUVILLE THEOREM. *Let u be a harmonic function on all of $\mathbf{E}^n$ and $u \geqq 0$. Then u is constant.*

We shall show that for each $\mathbf{y}$ in $\mathbf{E}^n$, $u(\mathbf{y}) = u(0)$. We fix $\mathbf{y}$ with $|\mathbf{y}| = b$ and select any $a > b$. Then

$$u(\mathbf{y}) = \frac{a^2-b^2}{aA_{n-1}}\int_{|\mathbf{x}|=a}\frac{u(\mathbf{x})}{|\mathbf{x}-\mathbf{y}|^n}\,\mu_{\mathbf{x}}.$$

Now

$$|\mathbf{x}-\mathbf{y}| \leqq |\mathbf{x}| + |\mathbf{y}| = a + b,$$

$$|\mathbf{x}-\mathbf{y}| \geqq |\mathbf{x}| - |\mathbf{y}| = a - b,$$

hence

$$\frac{1}{(a+b)^n} \leqq \frac{1}{|\mathbf{x}-\mathbf{y}|^n} \leqq \frac{1}{(a-b)^n}.$$

Since $u(\mathbf{x}) \geqq 0$ we may use these inequalities to estimate the integral:

$$\frac{(a^2-b^2)}{aA_{n-1}(a+b)^n}\int_{|\mathbf{x}|=a} u(\mathbf{x})\,\mu_{\mathbf{x}} \leqq u(\mathbf{y}) \leqq \frac{(a^2-b^2)}{aA_{n-1}(a-b)^n}\int_{|\mathbf{x}|=a} u(\mathbf{x})\,\mu_{\mathbf{x}}.$$

But from the Mean Value Theorem,

$$\int_{|\mathbf{x}|=a} u(\mathbf{x})\,\mu_{\mathbf{x}} = (a^{n-1}A_{n-1})\,u(0),$$

so we have

$$\frac{(a^2-b^2)a^{n-2}}{(a+b)^n}u(0) \leqq u(\mathbf{y}) \leqq \frac{(a^2-b^2)a^{n-2}}{(a-b)^n}u(0).$$

By letting $a \longrightarrow \infty$ we have

$$u(0) \leqq u(\mathbf{y}) \leqq u(0),$$

and so for each $\mathbf{y}$, $u(\mathbf{y}) = u(0)$, u is constant.

Remark 1. We return to the symmetrical Green's formula

$$\int_{\partial \mathbf{R}} (u *dv - v *du) = \int_{\mathbf{R}} (u\,\Delta v - v\,\Delta u)\,\omega.$$

We apply this in this situation:

$$\mathbf{R} = \{\varepsilon \leqq r \leqq a\},$$

v harmonic in $\mathbf{R}$ for all $\varepsilon > 0$,
v vanishes on $r = a$,
u smooth in $\{r \leqq a\}$.

We do not require that u be harmonic. The formula reduces to

$$\int_{r=a} u *dv - \int_{r=\varepsilon} u *dv + \int_{r=\varepsilon} v *du + \int_{\varepsilon \leqq r \leqq a} v(\Delta u)\,\omega = 0.$$

First case.

$$v = \frac{1}{r^{n-2}} - \frac{1}{a^{n-2}},$$

$$*dv = -(n-2)\tau.$$

By the methods above,

$$\int_{r=\varepsilon} u *dv = \frac{-(n-2)}{\varepsilon^n} \int_{r=\varepsilon} u\sigma = -(n-2)\,A_{n-1}u(0) + O(\varepsilon),$$

$$\int_{r=\varepsilon} v *du = \left(\frac{1}{\varepsilon^{n-2}} - \frac{1}{a^{n-2}}\right) \int_{r \leqq \varepsilon} d(*du) = O(\varepsilon^2).$$

Substituting these in and letting $\varepsilon \longrightarrow 0$ we obtain

$$u(0) = \frac{1}{a^{n-1}A_{n-1}} \int_{r=a} u\,\mu_{\mathbf{x}} - \frac{1}{(n-2)\,A_{n-1}} \int_{r \leqq a} \left(\frac{1}{r^{n-2}} - \frac{1}{a^{n-2}}\right)(\Delta u)\,\omega.$$

This gives information about a solution u of the Poisson equation $\Delta u = f$ with boundary values of u assigned.

Second case.

$$v = \frac{x_i}{r^n} - \frac{x_i}{a^n},$$

$$*dv = \left(\frac{1}{r^n} - \frac{1}{a^n}\right) *dx_i - n\,\frac{x_i}{r^{n+2}}\,\sigma.$$

One differentiates this to prove $\Delta v = 0$. This also follows when one notes that

$$\frac{\partial}{\partial x^i}\left(\frac{1}{r^{n-2}}\right) = \frac{-(n-2)}{r^n}\,x_i.$$

The end result in this case is

$$\left.\frac{\partial u}{\partial x_i}\right|_0 = \frac{n}{a^{n+1}A_{n-1}}\int_{r=a} x_i\, u\, \mu - \frac{1}{A_{n-1}}\int_{r\leqq a} x_i\left(\frac{1}{r^n} - \frac{1}{a^n}\right)(\Delta u)\,\omega.$$

Analogous formulas for higher derivatives are possible.

Remark 2. We have avoided $n = 2$. In this case the basic difference is that the symmetrical harmonic function with singularity at 0 is $\ln r$ rather than $r^{-(n-2)}$. Using this, results similar to those above follow.

7.2. The Heat Equation

We consider the parabolic equation

$$\frac{\partial^2 u}{\partial x^2} + \frac{\partial^2 u}{\partial y^2} = \frac{\partial u}{\partial t}\,.$$

Suppose u is a solution, valid in a region of x, y, t space which includes a region $\mathbf{R}$ and its boundary.

First we consider

$$\alpha = (u_x\, dy - u_y\, dx)\, dt - u\, dx\, dy.$$

Then

$$d\alpha = (u_{xx} + u_{yy})\, dx\, dy\, dt - u_t\, dt\, dx\, dy = 0,$$

hence

$$\int_{\partial\mathbf{R}} \alpha = \int_{\mathbf{R}} d\alpha = 0.$$

Next we consider

$$\beta = 2u(u_x\, dy - u_y\, dx)\, dt - u^2\, dx\, dy.$$

Then

$$\begin{aligned} d\beta &= 2(u_x\, dx + u_y\, dy)(u_x\, dy - u_y\, dx)\, dt + 2u(u_{xx} + u_{yy})\, dx\, dy\, dt - 2uu_t\, dt\, dx\, dy \\ &= 2(u_x^2 + u_y^2)\, dx\, dy\, dt. \end{aligned}$$

It follows that

$$\int_{\partial\mathbf{R}} [2u(u_x\, dy - u_y\, dx)\, dt - u^2\, dx\, dy] = 2\int_{\mathbf{R}} (u_x^2 + u_y^2)\, dx\, dy\, dt.$$

Suppose $\mathbf{R}$ is taken in the special form of a cylinder $\mathbf{T} \times [0, b]$ where $\mathbf{T}$ is a region in the x, y-plane. We then have

$$\partial\mathbf{R} = (\partial\mathbf{T}) \times [0, b] + \mathbf{T} \times \{b\} - \mathbf{T} \times \{0\}.$$

We now assert the basic uniqueness theorem:

If u vanishes on the base $\mathbf{T} \times \{0\}$ and on the lateral surface $\partial\mathbf{T} \times [0, b]$, then u vanishes identically in $\mathbf{R}$. For the integral formula above reduces to

$$2\int_{\mathbf{R}} (u_x^2 + u_y^2)\, dx\, dy\, dt + \int_{\mathbf{T}} u(x, y, b)^2\, dx\, dy = 0.$$

Since everything is positive, this implies

$$\begin{cases} u_x = u_y = 0 & \text{in } \mathbf{R}, \\ u = 0 & \text{on } \mathbf{T} \times \{b\}, \end{cases}$$

which is more than enough to imply $u = 0$ in $\mathbf{R}$. Because the heat equation is linear, we deduce that two temperature distributions which coincide initially at $t = 0$ and always coincide on the boundary of $\mathbf{R}$ must be identical for all t at each point of $\mathbf{R}$.

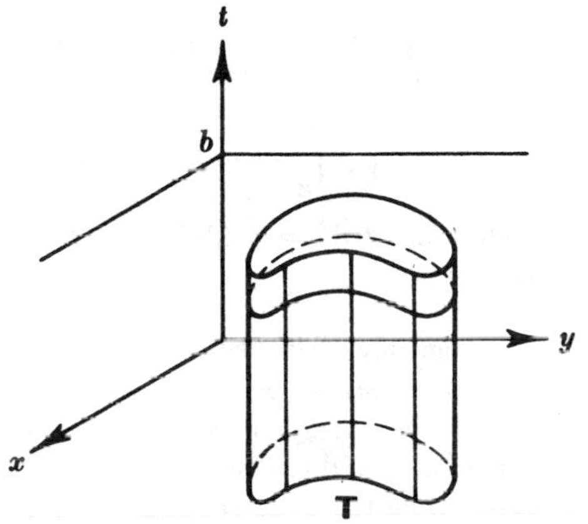

We shall now do the same thing in n dimensions, where there is an interesting sign change. Our variables are $x_1, \cdots, x_n, t$ and the heat equation is

$$\Delta u = \partial u/\partial t$$

where as usual $\Delta u = \sum \partial^2 u/\partial x_i^2$. The operator $*$ will apply to space variables only.

This time we set

$$\beta = 2u(*du)\,dt + (-1)^{n-1}u^2\omega,$$

where $\omega = dx_1 \cdots dx_n$. Now

$$du \wedge (*du) = (\text{grad}\, u)^2\omega = \sum \left(\frac{\partial u}{\partial x_i}\right)^2 \omega$$

and we have

$$\begin{aligned} d\beta &= 2(\text{grad}\, u)^2\omega\, dt + 2u(\Delta u)\,\omega\, dt + 2(-1)^{n-1}uu_t\, dt\,\omega \\ &= 2(\text{grad}\, u)^2\omega\, dt, \end{aligned}$$

hence

$$\int_{\partial\mathbf{R}} \beta = 2\int_{\mathbf{R}} (\text{grad}\, u)^2\omega\, dt.$$

Now let $\mathbf{T}$ be a region in $\mathbf{E}^n$, $\mathbf{R} = \mathbf{T} \times [0, b]$. Then

$$\begin{aligned}\partial\mathbf{R} &= \partial\mathbf{T} \times [0, b] + (-1)^n\mathbf{T} \times \partial[0, b] \\ &= \partial\mathbf{T} \times [0, b] + (-1)^n\mathbf{T} \times b + (-1)^{n-1}\mathbf{T} \times 0.\end{aligned}$$

Suppose u vanishes on $\partial\mathbf{T} \times [0, b]$ and on $\mathbf{T} \times 0$. Since $dt = 0$ on $\mathbf{T} \times b$ (i.e., $t = b =$ constant) we have

$$\int_{\partial\mathbf{R}} u *du\, dt = 0,$$

consequently

$$(-1)^{n-1}\int_{\partial\mathbf{R}} u^2\omega = 2\int_{\mathbf{R}} (\operatorname{grad} u)^2\omega\, dt,$$

that is

$$\int_{\mathbf{T}\times b} u^2\omega + 2\int_{\mathbf{R}} (\operatorname{grad} u)^2\omega\, dt = 0$$

and we conclude as before $(\operatorname{grad} u)^2 = \sum (\partial u/\partial x_i)^2 = 0$ on $\mathbf{R}$, $\partial u/\partial x_i = 0$ on $\mathbf{R}$, u is constant on $\mathbf{R}$, $u = 0$.

7.3. The Frobenius Integration Theorem†

Everything is local in this section; we operate in a neighborhood of 0 in $\mathbf{E}^n$. Let ω be a one-form which does not vanish at 0. We ask, under what conditions are there functions f and g satisfying $\omega = f\,dg$? In other words, we seek an integrating factor for the differential equation $\omega = 0$. If $\omega = f\,dg$, then f does not vanish in a neighborhood of 0, hence

$$d\omega = df \wedge dg = df \wedge f^{-1}\omega,$$

$$d\omega = \theta \wedge \omega \qquad (\theta = f^{-1}\,df = d\ln|f|)$$

and so

$$\omega \wedge d\omega = \omega \wedge \theta \wedge \omega = 0.$$

For a one-form $\omega = P\,dx + Q\,dy + R\,dz$ in $\mathbf{E}^3$, this is the condition

$$P(R_y - Q_z) + Q(P_z - R_x) + R(Q_x - P_y) = 0.$$

We note that if $\omega = f\,dg$, then the equations $\omega = 0$ and $dg = 0$ are the same and hence the solutions or *integral surfaces* of $\omega = 0$ are the hypersurfaces $g =$ constant.

Before passing on to precise statements we give two instructive examples.

Example 1. Let $\omega = yz\,dx + xz\,dy + dz$ so that $d\omega = y\,dz\,dx + x\,dz\,dy$. It follows that

$$d\omega = \left(\frac{dz}{z}\right) \wedge \omega$$

† Material in this and the next section is taken from a University of California Technical Report of June 1957, *Seminar on exterior differential forms*. This was prepared for the U.S. Army Office of Ordnance Research Contract DA–04–200–ORD–456.

which is not so useful since dz/z is singular along the z-axis. A better choice is $\theta = -y\,dx - x\,dy$ and we have $d\omega = \theta \wedge \omega$. To determine the function g, we use the fact that each integral surface $g =$ constant will be cut by the plane $\{x = at,\ y = bt\}$ in a curve which intersects the z-axis in the solution z of $g(0, 0, z) =$ constant. The equation $\omega = 0$ on the plane $x = at,\ y = bt$ becomes

$$dz + 2abzt\,dt = 0$$

with solution

$$z = c\exp(-abt^2)$$

satisfying the initial condition $z(0) = c$. However $abt^2 = xy$ so these curves span out a surface

$$z = ce^{-xy}.$$

We now think of a, b, c as variables and make the transformation

$$\begin{cases} x = a \\ y = b \\ z = ce^{-ab} \end{cases} \qquad \text{with} \qquad \frac{\partial(x, y, z)}{\partial(a, b, c)} = e^{-ab} \neq 0.$$

We have

$$dz = e^{-ab}dc - z(a\,db + b\,da),$$

which yields

$$\omega = e^{-ab}\,dc,$$

or in the original variables

$$\omega = e^{-xy}\,d(ze^{xy})$$

and the integral surfaces are

$$ze^{xy} = \text{constant}.$$

It will be observed that we have arranged the function g so that $g = c$ intersects the z-axis precisely in $z = c$.

Example 2. This time we try the procedure on $\omega = dz - y\,dx - dy$. On the plane $x = at,\ y = bt$, the equation $\omega = 0$ becomes $dz = (abt + b)\,dt$, $z = \frac{1}{2}abt^2 + bt + c$ and we arrive at the surface

$$z = \tfrac{1}{2}xy + y + c.$$

But on the parabolic cylinders $x = at,\ y = bt^2$ we have

$$dz = (abt^2 + 2bt)\,dt, \qquad z = \tfrac{1}{3}abt^3 + bt^2 + c,$$

$$z = \tfrac{1}{3}xy + y + c,$$

a different family of surfaces. The reason for this failure to obtain integral surfaces is seen from

$$d\omega = -dy\,dx, \qquad \omega \wedge d\omega = -dz\,dy\,dx \neq 0.$$

THEOREM. *Let $\omega = \sum f_i dx^i$ be a one-form which does not vanish at* 0. *Suppose there is a one-form θ satisfying $d\omega = \theta \wedge \omega$. Then there are functions f and g in a sufficiently small neighborhood of* 0 *which satisfy $\omega = f\,dg$.*

Note 1. Since ω is given while θ is certainly not uniquely determined, it simplifies the proof if we avoid explicit use of θ as long as possible.

Note 2. The condition on ω is unchanged when we replace ω by a multiple of ω. In fact, if $h \neq 0$, then

$$d(h\omega) = dh \wedge \omega + h\,d\omega = dh \wedge \omega + h\theta \wedge \omega = (dh + h\theta) \wedge \omega,$$

$$d(h\omega) = [(dh)h^{-1} + \theta](h\omega).$$

Proof. Since $\omega \neq 0$ at 0, we may assume some one of the functions f_i does not vanish at 0. Since neither the hypothesis nor conclusion changes when we multiply ω by a nonvanishing factor, we may assume

$$\omega = dz - \sum_1^n A_i dx^i, \qquad A_i = A_i(\mathbf{x}, z).$$

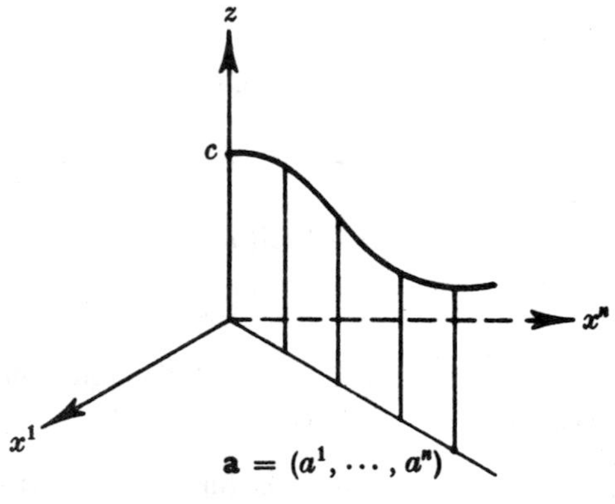

We fix any point $\mathbf{a}$ in $\mathbf{x}$-space and consider the equation $\omega = 0$ on the hyperplane $x^i = a^i t$, $(i = 1, \cdots, n)$:

$$\frac{dz}{dt} = \sum A_i(\mathbf{a}t, z)a^i.$$

We solve this equation with the initial condition $z(0) = c$. More precisely, we seek a function $F(t, \mathbf{a}, c)$ satisfying

$$\begin{cases} F_t(t, \mathbf{a}, c) = \sum A_i[\mathbf{a}t, F(t, \mathbf{a}, c)]a^i \\ F(0, \mathbf{a}, c) = c. \end{cases}$$

The usual existence theorem of ordinary differential equations yields a unique solution. We see that a change of scale is possible:

$$F(t, \mathbf{a}, c) = F\left(kt, \frac{1}{k}\mathbf{a}, c\right)$$

since the function on the right is again a solution to the same problem. In particular, setting $k = 1/t$,

$$F(t, \mathbf{a}, c) = F(1, t\mathbf{a}, c).$$

We introduce the change of variables

$$\begin{cases} \mathbf{x} = \mathbf{u} \\ z = F(1, \mathbf{u}, v) \end{cases}$$

with

$$\left.\frac{\partial(\mathbf{x}, z)}{\partial(\mathbf{u}, v)}\right|_0 = \begin{vmatrix} I & 0 \\ * & 1 \end{vmatrix} = 1$$

since

$$\left.\frac{\partial}{\partial v} F(1, \mathbf{u}, v)\right|_0 = \left.\frac{d}{dv} F(1, 0, v)\right|_{v=0} = \left.\frac{d}{dv} F(0, \mathbf{a}, v)\right|_{v=0}$$

$$= \left.\frac{dv}{dv}\right|_{v=0} = 1.$$

Thus the new variables $\mathbf{u}$, v form a local coordinate system in a sufficiently small neighborhood of 0. We suppose that in these coordinates we have

$$\omega = \sum P_i\, du^i + B\, dv, \qquad P_i = P_i(\mathbf{u}, v), \qquad B = B(\mathbf{u}, v).$$

Since ω vanishes identically on $\mathbf{u} = \mathbf{a}t$, $v =$ constant, we have the relation

$$\sum P_i(\mathbf{a}t, v)\, a^i = 0.$$

To continue, we consider the mapping ϕ on $(t, \mathbf{a}, v)$-space to $(\mathbf{u}, v)$-space given by

$$\phi(t, \mathbf{a}, v) = (t\mathbf{a}, v) = (\mathbf{u}, v).$$

We have

$$\begin{aligned} \phi^* \omega &= \sum P_i(t\mathbf{a}, v)(a^i\, dt + t\, da^i) + B(t\mathbf{a}, v)\, dv \\ &= \sum t P_i(t\mathbf{a}, v)\, da^i + B(t\mathbf{a}, v)\, dv, \\ \phi^* \omega &= \sum \bar{P}_i(t, \mathbf{a}, v)\, da^i + \bar{B}(t, \mathbf{a}, v)\, dv, \end{aligned}$$

where $\bar{P}_i(t, \mathbf{a}, v) = tP_i(t\mathbf{a}, v)$ so that $\bar{P}_i(0, \mathbf{a}, v) = 0$. The important point is that $\phi^* \omega$ is free of dt.

The equation $d\omega = \theta \wedge \omega$ implies $d(\phi^* \omega) = (\phi^* \theta) \wedge (\phi^* \omega)$. We may set

$$\phi^* \theta = H(t, \mathbf{a}, v)\, dt + \text{other terms}.$$

Then

$$d(\phi^* \omega) = \sum \frac{\partial \overline{P}_i}{\partial t} \, dt \, da^i + \text{other terms.}$$

We compare the $dt\,da^i$ terms in this on the one hand and $(\phi^* \theta) \wedge (\phi^* \omega)$ on the other to obtain

$$\frac{\partial \overline{P}^i}{\partial t} = H\overline{P}_i.$$

But this combined with $\overline{P}_i(0, \mathbf{a}, v) = 0$ implies, by the uniqueness theorem for ordinary equations, that $\overline{P}_i = 0$. Hence $P_i = 0$,

$$\omega = B\,dv,$$

the desired result.

References. For this theorem and the generalization which follows, see É. Cartan [9, p. 46], [7, p. 367].

Example. $\omega = x\,dy - y\,dx$. Certainly $\omega \wedge d\omega = 0$ since $\omega \wedge d\omega$ is a three-form. However, the form ω vanishes at 0 so one does not expect that the integral curves of $\omega = 0$ will span out evenly a neighborhood of 0; in fact these curves are just the lines $ax + by = 0$ through 0. We note, however, that $d\omega = \theta \wedge \omega$ is impossible in any neighborhood of 0. For $d\omega = 2\,dx\,dy$ so that if $\theta = A\,dx + B\,dy$, then $2 = Ax + By$ which fails at $x = y = 0$.

Remark. From the theorem we easily deduce again that a one-form ω satisfying $d\omega = 0$ is exact. For consider $\theta = dz - \omega$ where ω is a form in $\mathbf{x}$-space. Then $d\theta = 0$, hence there is a one-parameter family $z = F(\mathbf{x}, c)$ of integral surfaces, $F(0, c) = c$. For each choice of c, θ vanishes on $z = F(\mathbf{x}, c)$, i.e., $\omega = dF$. (We cannot proceed without passing to one more dimension since ω may vanish at 0.) This trick of introducing a new independent variable for an unknown function is a useful one.

We now pass to the general problem. Let $\omega^1, \cdots, \omega^r$ be one-forms in $r + s$ space, linearly independent at 0. Set $\Omega = \omega^1 \wedge \cdots \wedge \omega^r$. The system is called *completely integrable* if it satisfies any of the conditions of the following lemma.

LEMMA. *The following conditions are equivalent:*

(i) *There exist one-forms $\theta^i{}_j$ satisfying*

$$d\omega^i = \sum_{j=1}^{n} \theta^i{}_j \wedge \omega^j \qquad (i = 1, \cdots, r) \qquad (n = r + s)$$

(ii) $d\omega^i \wedge \Omega = 0 \qquad (i = 1, \cdots, r)$

(iii) *There exists a one-form λ satisfying*

$$d\Omega = \lambda \wedge \Omega.$$

Proof. That (i) implies (ii) is obvious (but unnecessary). Also (i) implies (iii) with $\lambda = \sum \theta^i{}_i$. Next, (iii) implies (ii) is the case since (iii) means

$$\sum (-1)^{i-1} d\omega^i \wedge \omega^1 \wedge \cdots \wedge \omega^{i-1} \wedge \omega^{i+1} \wedge \cdots \omega^r$$
$$= \lambda \wedge \omega^1 \wedge \cdots \wedge \omega^r$$

and we merely multiply by ω^i to deduce (ii).

It remains to prove that (ii) implies (i). Let $\omega^{r+1}, \cdots, \omega^n$ be one-forms so that $\omega^1, \cdots, \omega^n$ form a basis of all one-forms. We write

$$d\omega^i = \sum_{j<k} f^i{}_{jk} \omega^j \wedge \omega^k.$$

Since $d\omega^i \wedge \Omega = 0$, we have

$$\sum_{r<j<k} f^i{}_{jk} \omega^1 \wedge \cdots \wedge \omega^r \wedge \omega^j \wedge \omega^k = 0,$$

hence $f^i{}_{jk} = 0$ for $r < j < k$,

$$d\omega^i = \sum_{j=1}^{r} \left(\sum_{k=j+1}^{n} - f^i{}_{jk} \omega^k \right) \wedge \omega^j.$$

FROBENIUS INTEGRATION THEOREM. *Let $\omega^1, \cdots, \omega^r$ be one-forms in $\mathbf{E}^n$, $n = r + s$, linearly independent at 0. Suppose there are one-forms $\theta^i{}_j$ satisfying*

$$d\omega^i = \sum_{j=1}^{r} \theta^i{}_j \wedge \omega^j \qquad (i = 1, \cdots, r).$$

Then there are functions $f^i{}_j$, g^j satisfying

$$\omega^i = \sum_{j=1}^{r} f^i{}_j dg^j \qquad (i = 1, \cdots, r).$$

Discussion. The hypothesis is certainly a necessary one. For if we write

$$\boldsymbol{\omega} = (\omega^1, \cdots, \omega^r), \qquad F = \|f^i{}_j\|, \qquad \mathbf{g} = (g^1, \cdots, g^r),$$

the conclusion is $\boldsymbol{\omega} = d\mathbf{g}\, F$. The matrix F must be nonsingular in a neighborhood of 0 and so

$$d\boldsymbol{\omega} = -d\mathbf{g} \wedge dF = -\boldsymbol{\omega} \wedge F^{-1} dF = \boldsymbol{\omega} \wedge \Theta$$

where

$$\Theta = -F^{-1} dF.$$

Next we note the hypothesis is invariant under a linear transformation of the ω^i. In fact, if $\boldsymbol{\eta} = \boldsymbol{\omega} A$ where A is an $r \times r$ matrix of functions, nonsingular near 0, then

$$d\boldsymbol{\eta} = d\boldsymbol{\omega}\, A + \boldsymbol{\omega} \wedge dA = \boldsymbol{\omega} \wedge \Theta A + \boldsymbol{\omega} \wedge dA$$
$$= \boldsymbol{\eta} \wedge (A^{-1} \Theta A + A^{-1} dA).$$

We shall give two proofs of the theorem, each from a somewhat different point of view. The starting point is always the same. We write

$$\omega^i = \sum_{j=1}^{n} h^i{}_j dx^j \qquad (i = 1, \cdots, r).$$

Since the ω^i are linearly independent at 0, some $r \times r$ minor of $\|h^i{}_j\|$ is nonsingular in a neighborhood of 0. We multiply $(\omega^1, \cdots, \omega^r)$ by the inverse of this minor. On changing our notation slightly then we have

$$\omega^i = dz^i - \sum_{j=1}^{s} A^i{}_j(x^1, \cdots, x^s, z^1, \cdots, z^r)\,dx^j \qquad (i = 1, 2, \cdots, r).$$

First proof. For each point $\mathbf{a} = (a^1, \cdots, a^s)$ in $\mathbf{x}$-space we consider the system of equations $\omega^i = 0$ along the linear variety $\mathbf{x} = t\mathbf{a}$:

$$\frac{dz^i}{dt} = \sum A^i{}_j(t\mathbf{a}, \mathbf{z})\,a^j$$

with initial conditions $z^i(0) = c^i$. By ordinary differential equations, there is a unique solution in a sufficiently small neighborhood of 0, i.e., there exist functions $F^i(t, \mathbf{a}, \mathbf{c})$ satisfying

$$\frac{\partial F^i}{\partial t}(t, \mathbf{a}, \mathbf{c}) = \sum_{j=1}^{s} A^i{}_j[t\mathbf{a}, \mathbf{F}(t, a, \mathbf{c})]\,a^j$$

$$F^i(0, \mathbf{a}, \mathbf{c}) = c^i \qquad (i = 1, \cdots, r).$$

We shall write $\mathbf{F} = (F^1, \cdots, F^r)$.

Next, we fix k and set $\mathbf{G}(t, \mathbf{a}, \mathbf{c}) = \mathbf{F}(kt, \mathbf{a}, \mathbf{c})$. Then $\mathbf{G}(0, \mathbf{a}, \mathbf{c}) = \mathbf{c}$ and

$$\frac{\partial G^i}{\partial t}(t, \mathbf{a}, \mathbf{c}) = k\frac{\partial F^i}{\partial t}(kt, \mathbf{a}, \mathbf{c}) = \sum A^i{}_j(tk\mathbf{a}, \mathbf{G})\,ka^j,$$

hence by uniqueness, $\mathbf{G}(t, \mathbf{a}, \mathbf{c}) = \mathbf{F}(t, k\mathbf{a}, \mathbf{c})$, i.e.,

$$\mathbf{F}(kt, \mathbf{a}, \mathbf{c}) = \mathbf{F}(t, k\mathbf{a}, \mathbf{c}).$$

In particular, setting $t = 1$ and then replacing k by t,

$$\mathbf{F}(t, \mathbf{a}, \mathbf{c}) = \mathbf{F}(1, t\mathbf{a}, \mathbf{c}).$$

We pass to new variables $\mathbf{u}$, $\mathbf{v}$ according to the transformation

$$\mathbf{x} = \mathbf{u}$$

$$\mathbf{z} = \mathbf{F}(1, \mathbf{u}, \mathbf{v}).$$

This is nonsingular in some neighborhood of 0 since

$$\left.\frac{\partial(\mathbf{x}, \mathbf{z})}{\partial(\mathbf{u}, \mathbf{v})}\right|_0 = \begin{vmatrix} I & 0 \\ * & \dfrac{\partial \mathbf{z}}{\partial \mathbf{v}} \end{vmatrix}_0 = \left|\frac{\partial \mathbf{z}}{\partial \mathbf{v}}\right|_0 = 1.$$

For

$$\left.\frac{\partial z^i}{\partial v^j}\right|_0 = \left.\frac{\partial}{\partial v^j} F^i(1, 0, \mathbf{v})\right|_0 = \left.\frac{\partial}{\partial v^j} F^i(0, \mathbf{a}, \mathbf{v})\right|_0 = \left.\frac{\partial v^i}{\partial v^j}\right|_0 = \delta^i_j .$$

In these new variables we may write

$$\omega^i = \sum_{k=1}^{r} B^i{}_k(\mathbf{u}, \mathbf{v})\, dv^k + \sum_{j=1}^{s} P^i{}_j(\mathbf{u}, \mathbf{v})\, du^j.$$

The fact that each ω^i vanishes identically along the curve $\mathbf{u} = t\mathbf{a}$, $\mathbf{v} =$ constant implies

$$\sum_{j=1}^{s} P^i{}_j(t\mathbf{a}, \mathbf{v}) a^j = 0 \qquad (i = 1, \cdots, r).$$

We propose to show that the functions $P^i{}_j(\mathbf{u}, \mathbf{v})$ vanish identically. To do this we consider the cone mapping ϕ on $(t, \mathbf{a}, \mathbf{v})$-space to $(\mathbf{u}, \mathbf{v})$-space defined by

$$\phi(t, \mathbf{a}, \mathbf{v}) = (t\mathbf{a}, \mathbf{v}) = (\mathbf{u}, \mathbf{v}).$$

We have

$$\begin{aligned} \phi^* \omega^i &= \sum P^i{}_j(t\mathbf{a}, \mathbf{v})\, t\, da^j + \text{terms in } dv^k \\ &= \sum \bar{P}^i{}_j(t, \mathbf{a}, \mathbf{v})\, da^j + \text{terms in } dv^k \end{aligned}$$

where $\bar{P}^i{}_j(t, \mathbf{a}, \mathbf{v}) = P^i{}_j(t\mathbf{a}, \mathbf{v})t$ so that $\bar{P}^i{}_j(0, \mathbf{a}, \mathbf{v}) = 0$. It follows that

$$d\, \phi^* \omega^i = \sum \frac{\partial \bar{P}^i{}_j}{\partial t}\, dt\, da^j + \text{other terms}.$$

Finally we use the hypotheses $d\omega^i = \sum \theta^i{}_k \wedge \omega^k$. We write

$$\phi^* \theta^i{}_k = H^i{}_k(t, \mathbf{a}, \mathbf{v})\, dt + \text{other terms}$$

and compare the coefficients of $dt\, da^j$ in the relation

$$d\, \phi^* \omega^i = \sum (\phi^* \theta^i{}_k) \wedge (\phi^* \omega^k):$$

$$\frac{\partial \bar{P}^i{}_j}{\partial t}(t, \mathbf{a}, \mathbf{v}) = \sum H^i{}_k(t, \mathbf{a}, \mathbf{v})\, \bar{P}^k{}_j(t, \mathbf{a}, \mathbf{v}).$$

We conclude from the uniqueness of solutions of ordinary systems together with the initial conditions $\bar{P}^i{}_j(0, \mathbf{a}, \mathbf{v}) = 0$ that $\bar{P}^i{}_j = 0$, $P^i{}_j = 0$,

$$\omega^i = \sum_{k=1}^{r} B^i{}_k(\mathbf{u}, \mathbf{v})\, dv^k$$

as required.

Our next proof is based on the sketch in É. Cartan [10, pp. 188, 193]. We begin as before with the system

$$\omega^i = dz^i - \sum_{j=1}^{s} A^i{}_j dx^j \qquad (i = 1, \cdots, r)$$

with the conditions $d\omega^i = \sum \theta^i{}_j \wedge \omega^j$. We take any smooth curve from the origin to a point $\mathbf{a}$. We solve the system $\omega^i = 0$ on the cylinder this curve spans in $\mathbf{x}$, $\mathbf{z}$-space, taking some definite initial point $\mathbf{c}$ on the $\mathbf{z}$-axis. We shall show that the point on this curve lying over $\mathbf{x} = \mathbf{a}$ is *independent of the particular curve we start with in* $\mathbf{x}$-space. The point is that in a sufficiently small neighborhood of 0 in $\mathbf{x}$-space, any two smooth curves with the same end points 0, $\mathbf{a}$ can be smoothly deformed, one to the other. Thus let

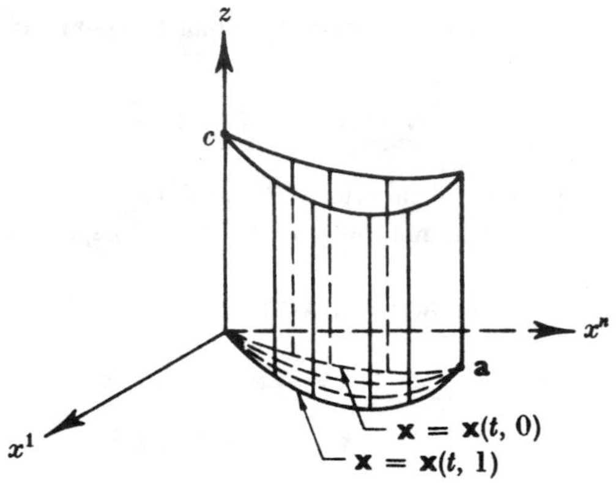

$\mathbf{x} = \mathbf{x}(t, \alpha)$ be a one-parameter family of curves from 0 to $\mathbf{a}$, the time variable t on each curve running from 0 to 1 and the parameter α taking all real values; we are interested in the curves $\mathbf{x}(t, 0)$ and $\mathbf{x}(t, 1)$. We are assuming

$$\mathbf{x}(0, \alpha) = 0, \qquad \mathbf{x}(1, \alpha) = \mathbf{a}.$$

Fixing α, the solution of $\omega^i = 0$ on the corresponding cylinder with initial value $\mathbf{c}$ is given by functions $F^i(t, \alpha)$ satisfying

$$\begin{cases} \dfrac{\partial F^i}{\partial t} = \displaystyle\sum_{j=1}^{s} A^i{}_j\big(\mathbf{x}(t, \alpha), \mathbf{F}(t, \alpha)\big) \dfrac{\partial x^j}{\partial t} \\ F^i(0, \alpha) = c^i \qquad (i = 1, \cdots, r). \end{cases}$$

We reduce the problem to a two-dimensional one by considering the mapping ϕ on (t, α)-space to $(\mathbf{x}, \mathbf{z})$-space given by

$$\phi(t, \alpha) = \big(\mathbf{x}(t, \alpha), \mathbf{F}(t, \alpha)\big) = (\mathbf{x}, \mathbf{z}).$$

Then

$$\begin{aligned} \phi^* \omega^i &= \frac{\partial F^i}{\partial t} dt + \frac{\partial F^i}{\partial \alpha} d\alpha - \sum A^i{}_j \frac{\partial x^j}{\partial t} dt - \sum A^i{}_j \frac{\partial x^j}{\partial \alpha} d\alpha \\ &= H^i \, d\alpha \end{aligned}$$

where

$$H^i = H^i(t, \alpha) = \frac{\partial F^i}{\partial \alpha} - \sum A^i{}_j \frac{\partial x^j}{\partial \alpha}.$$

We set

$$\phi^* \theta^i{}_j = P^i{}_j dt + Q^i{}_j d\alpha$$

and compare coefficients of $dt\, d\alpha$ in

$$d\,\phi^* \omega^i = \sum (\phi^* \theta^i{}_j) \wedge (\phi^* \omega^j)$$

to obtain

$$\frac{\partial H^i}{\partial t} = \sum P^i{}_j H^j.$$

But

$$H^i(0, \alpha) = \left.\frac{\partial F^i}{\partial \alpha}\right|_{t=0} - \sum A^i{}_j(0, \alpha) \left.\frac{\partial x^j}{\partial \alpha}\right|_{t=0}$$

$$= \frac{d}{d\alpha} F^i(0, \alpha) \quad - \sum A^i{}_j(0, \alpha) \frac{dx^j(0, \alpha)}{d\alpha}$$

$$= \frac{d}{d\alpha} c^i - \sum A^i{}_j(0, \alpha) \frac{d0}{d\alpha} = 0.$$

It follows that $H^i = 0$,

$$\frac{\partial F^i}{\partial \alpha}(t, \alpha) = \sum A^i{}_j \frac{\partial x^j}{\partial \alpha}.$$

We apply this in particular at $t = 1$; here

$$\left.\frac{\partial x^j}{\partial \alpha}(t, \alpha)\right|_{t=1} = \frac{d}{d\alpha} x^j(1, \alpha) = \frac{d}{d\alpha} a^j = 0,$$

hence

$$\frac{\partial F^i}{\partial \alpha}(1, \alpha) = 0, \qquad F^i(1, \alpha) = \text{constant}.$$

We next fix the notation more precisely; we write $F^i(t, \alpha; \mathbf{a}, \mathbf{c})$ instead of $F^i(t, \alpha)$ so as to specify the dependence of this function on the initial conditions. Since $F^i(1, \alpha)$ is independent of α we may set

$$G^i(\mathbf{a}, \mathbf{c}) = F^i(1, \alpha; \mathbf{a}, \mathbf{c}),$$

and also

$$\mathbf{F} = (F^1, \cdots, F^r), \qquad \mathbf{G} = (G^1, \cdots, G^r).$$

Then we have the following facts:

(i) $\mathbf{G}(0, \mathbf{c}) = \mathbf{c}$

(ii) For fixed $\mathbf{c}$, $\mathbf{a} \longleftrightarrow \big(\mathbf{a}, \mathbf{G}(\mathbf{a}, \mathbf{c})\big)$ is a 1-1 correspondence on a neighborhood of 0 in $\mathbf{a}$-space onto a manifold $\mathbf{V}_{\mathbf{c}}$ in $\mathbf{a}$, $\mathbf{c}$-space. (For we simply take any curve from 0 to $\mathbf{a}$ and use it to define $\mathbf{F}$ and then $\mathbf{G}$.)

(iii) Each ω^i vanishes identically on $\mathbf{V_c}$. (For ω^i vanishes on each curve on $\mathbf{V_c}$, since it is a one-form it vanishes identically.)

We consider the mapping

$$(\mathbf{a}, \mathbf{c}) \longrightarrow (\mathbf{a}, \mathbf{G}(\mathbf{a}, \mathbf{c})) = (\mathbf{x}, \mathbf{z})$$

on $\mathbf{a}$, $\mathbf{c}$-space to $\mathbf{x}$, $\mathbf{z}$-space. Because of (ii),

$$\left|\frac{\partial(\mathbf{x}, \mathbf{z})}{\partial(\mathbf{a}, \mathbf{c})}\right|_0 = 1,$$

hence we may use $(\mathbf{a}, \mathbf{c})$ for a new coordinate system in some neighborhood of 0. Writing ω^i in these new coordinates and using (iii) shows us that ω^i involves only the differentials $dc^1, \cdots, dc^r$, which completes the proof.

The striking feature of these proofs is that we reduce the original system of partial differential equations (with integrability conditions) to a system of ordinary differential equations.

7.4. Applications of the Frobenius Theorem

Example 1: We begin with a question in matrix form which is motivated by considerations in differential geometry centering around infinitesimal transformations.

Let $\Omega = \|\omega_i{}^j\|$ be an $r \times r$ matrix of one-forms defined in a neighborhood of 0, say, in $\mathbf{E}^n$. We ask when it is possible to find an $r \times r$ matrix A of functions, nonsingular, satisfying

$$\Omega = dA\, A^{-1}.$$

To fix matters, let us require the initial value $A_0 = I$. It is convenient to set

$$\Theta = d\Omega - \Omega^2.$$

Then the basic result is this.

There is a matrix of functions A defined in a neighborhood of 0 such that both $A_0 = I$ and

$$\Omega = (dA)\, A^{-1}$$

if and only if

$$\Theta = 0.$$

When this is the case, then there is only one such matrix A. First of all suppose there is a solution A. Then $(dA)\, A^{-1} = \Omega$, $dA = \Omega A$, and we have

$$0 = d(dA) = d\Omega\, A - \Omega\, dA = (\Theta + \Omega^2)\, A - \Omega(\Omega A) = \Theta A.$$

Hence $\Theta A = 0$. Since A is nonsingular we have $\Theta = 0$. If B is another solution so that $dA = \Omega A$, $dB = \Omega B$ and $A_0 = B_0 = I$, then

$$\begin{aligned} d(B^{-1}A) &= -B^{-1}\, dB\, B^{-1}A + B^{-1}\, dA \\ &= -B^{-1}(\Omega B)\, B^{-1}A + B^{-1}(\Omega A) = 0, \end{aligned}$$

hence $B^{-1}A$ is constant, $B^{-1}A = (B^{-1}A)_0 = I$, $B = A$.

Now we come to the existence. We pass to $(n + r^2)$-dimensional space with coordinates $x^1, \cdots, x^n, z_i^j$ $(1 \leqq i, j \leqq r)$ and introduce the r^2 forms which are the coefficients of the matrix

$$\Lambda = dZ - \Omega Z, \qquad Z = \|z_i^j\|.$$

We are assuming $\Theta = 0$, hence we have

$$d\Lambda = -d\Omega\, Z + \Omega\, dZ = -\Omega^2 Z + \Omega(\Lambda + \Omega Z),$$

$$d\Lambda = \Omega\Lambda.$$

It follows that our system Λ, which is already in standard form, is completely integrable, hence there exists a matrix A of functions of $\mathbf{x}$ with prescribed initial values at $\mathbf{x} = 0$, so that $Z = A$ is an integral manifold of $\Lambda = 0$, that is,

$$dA = \Omega A.$$

We remark that if Ω is skew-symmetric, then A is orthogonal (provided the initial condition is $A_0 = I$). For we set $B = {}^tA^{-1}$, the inverse transpose of A, and have

$$B_0 = I, \qquad dB = -B(d\,{}^tA)\,B = -B\,{}^tA\,{}^t\Omega B = +\,\Omega B.$$

It follows by uniqueness that $B = A$.

Example 2. We next consider another equation:

$$dA = \Omega A - A\Omega.$$

Here Ω is the same as before, an $r \times r$ matrix of one-forms in a neighborhood of $\mathbf{E}^n$ and A is the unknown matrix of functions. Again we set $\Theta = d\Omega - \Omega^2$. Let us pose the problem this way. Can we find a solution A of the above equation taking an arbitrary initial value A_0? We seek a necessary condition by differentiating:

$$\begin{aligned} 0 = d(dA) &= d\Omega\, A - \Omega\, dA - dA\, \Omega - A\, d\Omega \\ &= (\Theta + \Omega^2)\, A - \Omega(\Omega A - A\Omega) - (\Omega A - A\Omega)\, \Omega - A(\Theta + \Omega^2), \end{aligned}$$

which simplifies to

$$\Theta A = A\Theta.$$

Since we are assuming the initial values A_0 may be arbitrarily prescribed, the values of A at each point of a sufficiently small neighborhood of 0 will fill an n-dimensional domain; the commutativity of Θ at such points with so many different A evidently implies that the matrix Θ of two-forms must be of type

$$\Theta = \alpha I$$

where α is a two-form. From

$$d\Omega - \Omega^2 = \alpha I$$

we have, differentiating,

$$\begin{aligned} d\alpha\, I &= -d\Omega\,\Omega + \Omega\, d\Omega \\ &= -(\alpha I + \Omega^2)\,\Omega + \Omega(\alpha I + \Omega^2) = 0. \end{aligned}$$

Thus, locally, $\alpha = d\sigma$ where σ is a one-form. The necessary condition we arrive at is this: There must exist a one-form σ satisfying

$$d\Omega - \Omega^2 = (d\sigma)\, I.$$

This can also be expressed another way: The matrix

$$H = \Omega - \sigma I$$

satisfies

$$dH - H^2 = 0.$$

Under this condition, the sufficiency is easily demonstrated. As before we form

$$\Gamma = dZ - \Omega Z + Z\Omega$$

in $\mathbf{x}$, $\mathbf{z}$-space, and note that

$$\begin{aligned} d\Gamma &= -d\Omega\, Z + \Omega\, dZ + dZ\,\Omega + Z\, d\Omega \\ &= -(\Omega^2 + \Theta)\, Z + \Omega(\Gamma + \Omega Z - Z\Omega) + (\Gamma + \Omega Z - Z\Omega)\,\Omega + Z(\Omega^2 + \Theta) \\ &= Z\Theta - \Theta Z + \Omega\Gamma + \Gamma\Omega, \end{aligned}$$

hence

$$d\Gamma = \Omega\Gamma + \Gamma\Omega,$$

which shows that the system Γ is a completely integrable one. The existence proof now proceeds as in the last example. Uniqueness can be handled this way. Since the system $\Gamma = 0$ is in normal form and is completely integrable we know there is a unique integral surface passing through a given initial point $Z|_0 = A_0$.

Example 3. We shall consider a type of system of partial differential equations known as a *system of A. Mayer* (see C. Carathéodory [6, pp. 26–31]).

We work in a neighborhood of 0 in $\mathbf{E}^{r+s}$ with coordinates $x^1, \cdots, x^s$, $z^1, \cdots, z^r$ as before and are given functions $B^i{}_j(\mathbf{x}, \mathbf{z})$, $i = 1, \cdots, r$, $j = 1, \cdots, s$. The Mayer system is

$$\frac{\partial z^i}{\partial x^j} = B^i{}_j(\mathbf{x}, \mathbf{z}).$$

We define

$$A^i{}_{jk} = \frac{\partial B^i{}_j}{\partial x^k} - \frac{\partial B^i{}_k}{\partial x^j} + \sum_\alpha \frac{\partial B^i{}_j}{\partial z^\alpha} B^\alpha{}_k - \sum_\beta \frac{\partial B^i{}_k}{\partial z^\beta} B^\beta{}_j .$$

We evidently have $A^i{}_{jk} + A^i{}_{kj} = 0$. The Mayer system is called *completely integrable* in a neighborhood of 0 provided to each choice of initial conditions $\mathbf{c}$ there exists a solution $\mathbf{z} = \mathbf{F}(\mathbf{x}, \mathbf{c})$ of the system with $\mathbf{F}(0, \mathbf{c}) = \mathbf{c}$. The necessary and sufficient condition for complete integrability is precisely $A^i{}_{jk} = 0$. The reason is the following. We set

$$\omega^i = dz^i - \sum_{j=1}^{s} B^i{}_j(\mathbf{x}, \mathbf{z})\, dx^j \qquad (i = 1, \cdots, r)$$

a system of one-forms in $\mathbf{x}$, $\mathbf{z}$-space in our standard form. The vanishing $A^i{}_{jk} = 0$ implies, after a short calculation,

$$d\omega^i = \sum \left(\sum \frac{\partial B^i{}_j}{\partial z^\alpha}\, dx^j \right) \wedge \omega^\alpha$$

so that the system $\omega^1, \cdots, \omega^r$ is completely integrable. The integral surfaces $\mathbf{z} = \mathbf{F}(\mathbf{x}, \mathbf{c})$ solve the Mayer system. Conversely, suppose the Mayer system is completely integrable. Then it is clear that the system $\omega^1 = \cdots = \omega^r = 0$ has integral surfaces, one for each choice of initial conditions $\mathbf{c}$. Hence the necessary condition

$$d\omega^i = \sum \theta^i{}_j \wedge \omega^j$$

must be satisfied. On the other hand, we directly verify that

$$d\omega^i = \tfrac{1}{2} \sum A^i{}_{jk}\, dx^j\, dx^k + \sum \eta^i{}_\alpha \wedge \omega^\alpha$$

where

$$\eta^i{}_\alpha = \sum \frac{\partial B^i{}_j}{\partial z^\alpha}\, dx^j.$$

Since $dx^1, \cdots, dx^s, \omega^1, \cdots, \omega^r$ are linearly independent we conclude that

$$\theta^i{}_j \wedge \omega^j = \sum \eta^i{}_\alpha \wedge \omega^\alpha, \qquad \sum A^i{}_{jk}\, dx^j\, dx^k = 0, \qquad A^i{}_{jk} = 0.$$

Note. If $\omega^1, \cdots, \omega^r$ is completely integrable, there is an $r \times r$ matrix of one-forms $\Theta = \|\theta^i{}_j\|$ satisfying

$$d\omega^i = \sum \theta^i{}_j \wedge \omega^j,$$

or

$$d\boldsymbol{\omega} = -\boldsymbol{\omega} \wedge \Theta$$

in matrix notation. However, from the solution

$$\boldsymbol{\omega} = d\mathbf{g}\, F, \qquad \mathbf{g} = (g^1, \cdots, g^r), \qquad F = \|f^i{}_j\|$$

we conclude that $d\boldsymbol{\omega} = -\boldsymbol{\omega} \wedge F^{-1}\, dF$ so that we may always choose Θ in the very special form $\Theta = F^{-1}\, dF$. (This provides some motivation for the problem in Example 1.)

We shall see further applications of the Frobenius theorem in our study of local Riemannian geometry. Also cf. Problems 4–7, p. 194.

7.5. Systems of Ordinary Equations

We consider a system

$$\frac{dx^1}{dt} = X^1(t, x^1, \cdots, x^n)$$

$$\cdots\cdots\cdots\cdots\cdots\cdots\cdots$$

$$\frac{dx^n}{dt} = X^n(t, x^1, \cdots, x^n).$$

Closely associated with this system is the differential n-form

$$\Omega = (dx^1 - X^1\,dt)\cdots(dx^n - X^n\,dt)$$

in $(t, \mathbf{x})$-space. By a short computation,

$$d\Omega = \left(\sum \frac{\partial X^i}{\partial x^i}\right) dt\,dx^1 \cdots dx^n.$$

We make a change of variables

$$y^i = y^i(t, x^1, \cdots, x^n) \qquad (i = 1, \cdots, n)$$

and suppose that the systems

$$\frac{dx^i}{dt} = X^i(t, \mathbf{x}) \qquad \text{and} \qquad \frac{dy^i}{dt} = Y^i(t, \mathbf{y})$$

are equivalent under this change. We set

$$\overline{\Omega} = (dy^1 - Y^1\,dt)\cdots(dy^n - Y^n\,dt)$$

and propose to determine how $d\Omega$ and $d\overline{\Omega}$ are related. Now

$$\frac{dy^i}{dt} = \frac{\partial y^i}{\partial t} + \sum \frac{\partial y^i}{\partial x^j}\frac{dx^j}{dt},$$

$$Y^i = \frac{\partial y^i}{\partial t} + \sum \frac{\partial y^i}{\partial x^j} X^j.$$

Also

$$dy^i = \frac{\partial y^i}{\partial t}\,dt + \sum \frac{\partial y^i}{\partial x^j}\,dx^j,$$

hence

$$dy^i - Y^i\,dt = \sum \frac{\partial y^i}{\partial x^j}(dx^j - X^j dt).$$

We denote the Jacobian by J:

$$J = \frac{\partial(y^1, \cdots, y^n)}{\partial(x^1, \cdots, x^n)} = \left|\frac{\partial y^i}{\partial x^j}\right|$$

and have by exterior multiplication,

$$\overline{\Omega} = J\Omega.$$

We differentiate this:

$$d\overline{\Omega} = (dJ)\,\Omega + J\,d\Omega.$$

Now

$$d\overline{\Omega} = \left(\sum \frac{\partial Y^i}{\partial y^i}\right) dt\,dy^1 \cdots dy^n = \left(J \sum \frac{\partial Y^i}{\partial y^i}\right) dt\,dx^1 \cdots dx^n$$

and

$$\begin{aligned}(dJ)\Omega + J\,d\Omega &= \left(\frac{\partial J}{\partial t}\,dt + \sum \frac{\partial J}{\partial x^i}\,dx^i\right)\Omega + J\left(\sum \frac{\partial X^i}{\partial x^i}\right) dt\,dx^1 \cdots dx^n \\ &= \left(\frac{\partial J}{\partial t} + \sum \frac{\partial J}{\partial x^i} X^i + J \sum \frac{\partial X^i}{\partial x^i}\right) dt\,dx^1 \cdots dx^n,\end{aligned}$$

hence

$$\sum \frac{\partial Y^i}{\partial y^i} = \frac{1}{J}\left(\frac{\partial J}{\partial t} + \sum \frac{\partial (JX^i)}{\partial x^i}\right).$$

A function $f = f(t, \mathbf{x})$ is called a *first integral* of the system if f is constant along each trajectory, or solution curve. Since the usual existence and uniqueness theorems guarantee a solution through each point of space where the system is defined, we have the condition

$$\frac{df}{dt} = 0,$$

i.e.,

$$\frac{df}{\partial t} + \sum \frac{\partial f}{\partial x^j} X^j = 0,$$

for a first integral.

Suppose that each of the functions $y^1, \cdots, y^n$ in the transformation above is a first integral. Then

$$Y^i = 0, \qquad \overline{\Omega} = dy^1 \cdots dy^n.$$

(Such a transformation is always possible in a small region of space because of the existence of a general solution, one depending on arbitrarily prescribed initial conditions.)

A function $M = M(t, x^1, \cdots, x^n)$ is called a *last multiplier* of the original system if

$$d(M\Omega) = 0,$$

i.e.,

$$\frac{\partial M}{\partial t} + \sum \frac{\partial (MX^i)}{\partial x^i} = 0.$$

Using the transformation above based on first integrals,

$$M\Omega = \left(\frac{M}{J}\right)\overline{\Omega} = H\overline{\Omega} = H\,dy^1 \cdots dy^n.$$

Hence

$$d(M\Omega) = (dH)\,dy^1 \cdots dy^n = \frac{\partial H}{\partial t}\,dt\,dy^1 \cdots dy^n$$

so that M is a last multiplier if and only if H is independent of t, i.e.,

$$M\Omega = H(\mathbf{y})\,dy^1 \cdots dy^n.$$

If M_1 and M_2 are two last multipliers, then

$$M_1\Omega = H_1(\mathbf{y})\,dy^1 \cdots dy^n, \qquad M_2\Omega = H_2(\mathbf{y})\,dy^1 \cdots dy^n,$$

hence

$$M_1/M_2 = H_1(\mathbf{y})/H_2(\mathbf{y}).$$

It follows that M_1/M_2 depends only on the y^i, hence is constant along trajectories (as is each y^i) and consequently is a first integral. This proves the important result:

The quotient of two last multipliers is a first integral.

7.6. The Third Lie Theorem

What is known as the Third Fundamental Theorem of Sophus Lie was devised in order to reconstruct a continuous group given only its constants of structure. These concepts will be explained in Chapter IX. For our present purposes, we shall look upon this theorem as a result, and a rather deep one at that, in partial differential equations.

We work in $\mathbf{E}^n$. All indices run from 1 to n. First of all we are given n^3 constants $c^i{}_{jk}$ subject to these constraints:

$$c^i{}_{jk} + c^i{}_{kj} = 0$$

$$\sum_j (c^i{}_{jk}c^j{}_{rs} + c^i{}_{jr}c^j{}_{sk} + c^i{}_{js}c^j{}_{kr}) = 0.$$

The problem is to find n one-forms $\sigma^1, \cdots, \sigma^n$ which are linearly independent on some neighborhood of 0 in $\mathbf{E}^n$ and which satisfy the relations

$$d\sigma^i = \tfrac{1}{2} \sum c^i{}_{jk}\sigma^j \wedge \sigma^k.$$

The Lie Theorem asserts that this can be done.

The quadratic relations we have assumed for the constants (c) are easily verified to be the same as $d(d\sigma^i) = 0$, assuming our problem is solved, hence they are necessary conditions. That they also are sufficient conditions will now be seen. The proof we give is based on those in Cartan [10, p. 239] and [7, pp. 280–283]. Because the proof is lengthy, we shall break it into several steps.

Step 1. We define an $n \times n$ matrix $F = [f^i{}_k]$ of homogeneous linear forms by

$$f^i{}_k = \sum c^i{}_{jk}\, x^j.$$

Then we consider the linear initial value problem

$$\frac{\partial H}{\partial t} = I + HF, \quad H(0, \mathbf{x}) = 0$$

for an $n\times n$ matrix $H = H(t, \mathbf{x})$. For each $\mathbf{x}$ it has a unique solution, defined for all t, and the solution is analytic in $(t, \mathbf{x})$. At $\mathbf{x} = \mathbf{0}$ we have $F = 0$, so

$$\frac{\partial H}{\partial t}(t, \mathbf{0}) = I, \quad H(0, \mathbf{0}) = 0,$$

hence

$$H(t, \mathbf{0}) = tI \quad \text{and} \quad H(1, \mathbf{0}) = I.$$

Step 2. We set

$$\boldsymbol{\omega} = d\mathbf{x}H(t, \mathbf{x})$$

so that $\boldsymbol{\omega}$ is a row vector of one-forms, free of dt. Clearly

$$\begin{aligned} d\boldsymbol{\omega} &= dt\ d\mathbf{x}(I + HF) + \boldsymbol{\lambda} \\ &= dt(d\mathbf{x} + \boldsymbol{\omega}F) + \boldsymbol{\lambda}, \end{aligned}$$

where $\boldsymbol{\lambda}$ is a row vector of two-forms, also free of dt. We also define a square matrix $A = [\alpha^i{}_j]$ of one-forms by

$$\alpha^i{}_j = \sum c^i{}_{jk}\omega^k.$$

We note the obvious relations:

$$\mathbf{x}A = \boldsymbol{\omega}F \quad \text{and} \quad d\mathbf{x}A = -\boldsymbol{\omega}dF.$$

We also note two less obvious relations:

$$d(\boldsymbol{\omega}A) = 2d\boldsymbol{\omega} \wedge A \quad \text{and} \quad 2\mathbf{x}A^2 = \boldsymbol{\omega}AF.$$

The first follows from the skew-symmetry of the constants $c^i{}_{jk}$ in their lower indices. The second follows from the quadratic relations on the c's. We multiply the quadratic relation (p. 108) by $x^r\omega^k\omega^s$ and sum:

$$\sum x^r\alpha^i{}_j\alpha^j{}_r + \sum f^i{}_j c^j{}_{ks}\omega^k\omega^s + \sum x^r\alpha^i{}_j\alpha^j{}_r = 0,$$

which in matrix language is $-\mathbf{x}A^2 + \boldsymbol{\omega}AF - \mathbf{x}A^2 = 0$.

Step 3. From the formula for $d\boldsymbol{\omega}$ that defines $\boldsymbol{\lambda}$, we have

$$d\boldsymbol{\lambda} = dt \wedge d(\boldsymbol{\omega}F) = dt(d\boldsymbol{\omega}F - \boldsymbol{\omega}dF)$$
$$= dt(\boldsymbol{\lambda}F - \boldsymbol{\omega}dF) = dt(\boldsymbol{\lambda}F + d\mathbf{x}A).$$

Step 4. We define

$$\boldsymbol{\theta} = \boldsymbol{\lambda} - \tfrac{1}{2}\boldsymbol{\omega}A,$$

a row vector of two-forms, free of dt. We shall prove the decisive formula

$$d\boldsymbol{\theta} = dt \wedge \boldsymbol{\theta}F - \boldsymbol{\lambda} \wedge A.$$

We have

$$d\boldsymbol{\theta} = d\boldsymbol{\lambda} - d\boldsymbol{\omega} \wedge A = dt(\boldsymbol{\lambda}F + d\mathbf{x} \wedge A) - [dt(d\mathbf{x} + \boldsymbol{\omega}F) + \boldsymbol{\lambda}]A$$
$$= dt(\boldsymbol{\lambda}F - \boldsymbol{\omega}FA) - \boldsymbol{\lambda} \wedge A$$
$$= dt(\boldsymbol{\lambda}F - \mathbf{x}A^2) - \boldsymbol{\lambda} \wedge A$$
$$= dt(\boldsymbol{\lambda}F - \tfrac{1}{2}\boldsymbol{\omega}AF) - \boldsymbol{\lambda} \wedge A$$
$$= dt \wedge \boldsymbol{\theta}F - \boldsymbol{\lambda} \wedge A.$$

Step 5. We shall now prove that

$$\theta^i = 0.$$

Since both the ω^j and the λ^i are free of dt, so is θ^i and we have

$$\theta^i = \tfrac{1}{2}\sum g^i{}_{jk}\,dx^j\,dx^k, \qquad g^i{}_{jk} = g^i{}_{jk}(t, \mathbf{x}).$$

We may even assert that

$$g^i{}_{jk}(0, \mathbf{x}) = 0.$$

For $h^i{}_j(0, \mathbf{x}) = 0$ which not only means that the coefficients of the ω^i vanish at $t = 0$, but also that the coefficients of

$$\lambda^i = \tfrac{1}{2}\sum\left(\frac{\partial h^i{}_j}{\partial x^k} - \frac{\partial h^i{}_k}{\partial x^j}\right)dx^k\,dx^j$$

vanish at $t = 0$ according to the very definition of partial derivative,

$$\left.\frac{\partial h^i{}_j}{\partial x^k}\right|_{t=0} = \frac{\partial h^i{}_j(0, \mathbf{x})}{\partial x^k} = 0.$$

From these facts, what we say about the initial values of the $g^i{}_{jk}$ follows.

The result of Step 4 implies

$$\frac{\partial g^i{}_{jk}}{\partial t} = \sum c^i{}_{rs}\, x^r\, g^s{}_{jk}\,.$$

This homogeneous linear system taken together with the initial conditions, the vanishing of the g's at $t = 0$, has the unique solution

$$g^i{}_{jk}(t, \mathbf{x}) = 0,$$

and so $\theta^i = 0$.

Step 6. Now we can wind up this story. Since $\theta^i = 0$ we have

$$\lambda^i = \tfrac{1}{2} \sum c^i{}_{jk}\, \omega^j \wedge \omega^k,$$
$$d\omega^i = \tfrac{1}{2} \sum c^i{}_{jk}\, \omega^j \wedge \omega^k + dt \wedge \alpha^i.$$

We consider this relation on the subspace $t = 1$.
Setting

$$\sigma^i = \omega^i|_{t=1} = \sum h^i{}_j(1, \mathbf{x})\, dx^j,$$

it becomes

$$d\sigma^i = \tfrac{1}{2} \sum c^i{}_{jk}\, \sigma^j \wedge \sigma^k.$$

Since $h^i{}_j(1, 0) = \delta^i_j$ (Step 1), the one-forms $\sigma^1, \cdots, \sigma^n$ are linearly independent at 0, which implies of course that they are linearly independent in some neighborhood of 0. The proof is complete.

VIII

Applications to Differential Geometry

8.1. Surfaces (Continued)

Everything in this section will be based on the local theory of Section 4.5. Now we have integration at our disposal and we shall discuss a few global results. Let Σ be a closed surface in $\mathbf{E}^3$. For $\mathbf{e}_3$ we take the outward drawn normal to Σ. The mapping

$$\mathbf{x} \longrightarrow \mathbf{e}_3$$

is a map on Σ to the unit sphere $\mathbf{S}^2$. As $\mathbf{x}$ varies over Σ, $\mathbf{e}_3$ varies over $\mathbf{S}^2$ a whole number of times, called the *degree* of the normal map (cf. Section 6.2). The element of area of the normal map is

$$\omega_1\omega_2 = K\sigma_1\sigma_2$$

since

$$d\mathbf{e}_3 = \omega_1\mathbf{e}_1 + \omega_2\mathbf{e}_2 .$$

Here K is the Gaussian curvature. Hence

$$\int_\Sigma K\sigma_1\sigma_2 = 4\pi n$$

where n is the degree. The factor 4π is simply the area of the unit sphere.

In particular, if Σ is a closed *convex* surface, then $\mathbf{e}_3$ covers $\mathbf{S}^2$ exactly once as $\mathbf{x}$ covers Σ, hence

$$\int_\Sigma K\sigma_1\sigma_2 = 4\pi$$

in this case.

After this, we shall limit our discussion to closed convex surfaces. Two important invariants are the total area

$$A = \int_\Sigma \sigma_1\sigma_2$$

and the integrated mean curvature

$$M = \int_\Sigma H\sigma_1\sigma_2 .$$

Given a closed convex surface Σ and a fixed positive number a, we form the surface Σ' *parallel to* Σ at distance a by marking off on the outward-drawn normal at each point $\mathbf{x}$ of Σ the distance a and taking the locus of all points so obtained. Thus the typical point on the parallel surface is

$$\mathbf{y} = \mathbf{x} + a\mathbf{e}_3$$

where $\mathbf{e}_3$ always denotes the normal at $\mathbf{x}$. We have

$$\begin{aligned} d\mathbf{y} &= d\mathbf{x} + a\,d\mathbf{e}_3 \\ &= (\sigma_1\mathbf{e}_1 + \sigma_2\mathbf{e}_2) + a(\omega_1\mathbf{e}_1 + \omega_2\mathbf{e}_2) \\ &= (\sigma_1 + a\omega_1)\mathbf{e}_1 + (\sigma_2 + a\omega_2)\mathbf{e}_2 . \end{aligned}$$

It follows that the normal to the parallel surface Σ' at $\mathbf{y}$ is again $\mathbf{e}_3$ and that $\mathbf{e}_1$ and $\mathbf{e}_2$ can be taken as a basis of the tangent space at $\mathbf{y}$. Thus we have

$$d\mathbf{y} = \tau_1\mathbf{e}_1 + \tau_2\mathbf{e}_2$$

with

$$\tau_1 = \sigma_1 + a\omega_1, \qquad \tau_2 = \sigma_2 + a\omega_2 .$$

It follows that the element of area of Σ' is

$$\begin{aligned} \tau_1\tau_2 &= (\sigma_1 + a\omega_1)(\sigma_2 + a\omega_2) \\ &= \sigma_1\sigma_2 + a(\sigma_1\omega_2 - \sigma_2\omega_1) + a^2\omega_1\omega_2 \\ &= (1 + 2aH + a^2K)\sigma_1\sigma_2 \end{aligned}$$

so that the total area of Σ' is

$$\begin{aligned} A' &= \int \tau_1\tau_2 = \int (1 + 2aH + a^2K)\sigma_1\sigma_2 , \\ A' &= A + 2aM + 4\pi a^2. \end{aligned}$$

(This formula can also be proved by first doing it for a polyhedron and then taking limits in an approximation of Σ by a sequence of polyhedra. If one examines what the formula means when Σ is a convex polyhedron, one will see that H measures dihedral angle and K vertex angle.)

By integrating with respect to a, one easily comes to a relation between the three-dimensional volumes V' and V enclosed by Σ' and Σ, respectively:

$$V' = V + aA + a^2M + \tfrac{4}{3}\pi a^3.$$

One can also verify the relations

$$\begin{aligned} M' &= M + 4\pi a, \\ H' &= \frac{H + aK}{1 + 2aH + a^2K}, \\ K' &= \frac{K}{1 + 2aH + a^2K}. \end{aligned}$$

We next introduce the *support function* of our closed convex surface $\mathbf{\Sigma}$. This is defined by

$$p = \mathbf{x} \cdot \mathbf{e}_3.$$

It is convenient to fix $\mathbf{\Sigma}$ in space so that the origin O is inside $\mathbf{\Sigma}$. Then we have $p > 0$ at each point of $\mathbf{\Sigma}$.

The following method will be used to obtain several identities. Let λ be any one-form on $\mathbf{\Sigma}$. Then

$$\int_{\mathbf{\Sigma}} d\lambda = 0.$$

For $\partial\mathbf{\Sigma} = 0$ and Stokes' theorem gives us

$$\int_{\mathbf{\Sigma}} d\lambda = \int_{\partial\mathbf{\Sigma}} \lambda = 0.$$

First we consider the form

$$\alpha = \mathbf{e}_3 \cdot (\mathbf{x} \times d\mathbf{x}).$$

Here

$$d\alpha = d\mathbf{e}_3 \cdot (\mathbf{x} \times d\mathbf{x}) + \mathbf{e}_3 \cdot (d\mathbf{x} \times d\mathbf{x}).$$

Now

$$\begin{aligned} d\mathbf{e}_3 \cdot (\mathbf{x} \times d\mathbf{x}) &= -\mathbf{x} \cdot (d\mathbf{e}_3 \times d\mathbf{x}) \\ &= -\mathbf{x} \cdot (d\mathbf{x} \times d\mathbf{e}_3) = -\mathbf{x} \cdot (2H\sigma_1\sigma_2\mathbf{e}_3) \\ &= -2H\sigma_1\sigma_2(\mathbf{x} \cdot \mathbf{e}_3) = -2pH\sigma_1\sigma_2 \end{aligned}$$

and

$$\begin{aligned} \mathbf{e}_3 \cdot (d\mathbf{x} \times d\mathbf{x}) &= \mathbf{e}_3 \cdot (2\sigma_1\sigma_2\mathbf{e}_3) \\ &= 2\sigma_1\sigma_2, \end{aligned}$$

so that

$$d\alpha = 2[\sigma_1\sigma_2 - pH\sigma_1\sigma_2].$$

Since the integral of $d\alpha$ is zero,

$$A = \int_{\mathbf{\Sigma}} \sigma_1\sigma_2 = \int_{\mathbf{\Sigma}} pH\sigma_1\sigma_2.$$

Next we set

$$\beta = \mathbf{x} \cdot (\mathbf{e}_3 \times d\mathbf{e}_3).$$

By a calculation similar to that used for α,

$$d\beta = 2[pK\sigma_1\sigma_2 - H\sigma_1\sigma_2]$$

so that

$$M = \int_{\mathbf{\Sigma}} H\sigma_1\sigma_2 = \int_{\mathbf{\Sigma}} pK\sigma_1\sigma_2.$$

Since we have found the integrals of H and K weighted by p it is also reasonable to seek the integral over Σ of the form $p\sigma_1\sigma_2$. We get this by starting with the vectorial area

$$(\sigma_1\sigma_2)\mathbf{e}_3 = \tfrac{1}{2}d\mathbf{x} \times d\mathbf{x} = (dy\,dz,\, dz\,dx,\, dx\,dy)$$

from which

$$\begin{aligned} p\sigma_1\sigma_2 &= (\mathbf{x}\cdot\mathbf{e}_3)(\sigma_1\sigma_2) = \mathbf{x}\cdot(dy\,dz,\, dz\,dx,\, dx\,dy) \\ &= x\,dy\,dz + y\,dz\,dx + z\,dx\,dy. \end{aligned}$$

Let $\mathbf{R}$ be the region of $\mathbf{E}^3$ bounded by the closed convex surface and let V denote its volume. Then

$$\begin{aligned} \int_{\Sigma} p\sigma_1\sigma_2 &= \int_{(\Sigma=\partial\mathbf{R})} (x\,dy\,dz + y\,dz\,dx + z\,dx\,dy) \\ &= \int_{\mathbf{R}} d(x\,dy\,dz + \cdots) = 3\int_{\mathbf{R}} dx\,dy\,dz = 3V, \end{aligned}$$

$$\tfrac{1}{3}\int_{\Sigma} p\sigma_1\sigma_2 = V.$$

We close this section with the following interesting theorem:

Let Σ be a closed convex surface of constant Gaussian curvature K. Then Σ is a sphere.

To prove this, we recall the relations

$$\begin{cases} d\mathbf{x} = \sigma_1\mathbf{e}_1 + \sigma_2\mathbf{e}_2 \\ d\mathbf{e}_3 = \omega_1\mathbf{e}_1 + \omega_2\mathbf{e}_2 \end{cases}$$

$$\begin{cases} \omega_1 = p\sigma_1 + q\sigma_2 \\ \omega_2 = q\sigma_1 + r\sigma_2 \end{cases}$$

$$\begin{cases} H = \tfrac{1}{2}(p + r) \\ K = pr - q^2 \end{cases}$$

developed in Section 4.5 for any moving frame. Since Σ is convex, the matrix

$$\begin{pmatrix} p & q \\ q & r \end{pmatrix}$$

is positive definite, $p > 0$, $r > 0$, $K > 0$. (See p. 120–121 in the next section for details.) Because of the arithmetic-geometric mean inequality we have

$$K = pr - q^2 \leqq pr \leqq [\tfrac{1}{2}(p + r)]^2 = H^2.$$

We also note that there can be equality, $K = H^2$, only when $q = 0$ and $p = r = H$, which implies $\omega_1 = H\sigma_1$, $\omega_2 = H\sigma_2$, $d\mathbf{e}_3 = H\,d\mathbf{x}$.

But

$$\iint \sigma_1\sigma_2 = \iint pH\sigma_1\sigma_2 \geqq \iint p\sqrt{K}\sigma_1\sigma_2 = \frac{1}{\sqrt{K}}\iint pK\sigma_1\sigma_2$$
$$= \frac{1}{\sqrt{K}}\iint H\sigma_1\sigma_2 \geqq \frac{1}{\sqrt{K}}\iint \sqrt{K}\sigma_1\sigma_2 = \iint \sigma_1\sigma_2,$$

where each integral is taken over $\boldsymbol{\Sigma}$ and we have exploited the hypothesis $K =$ constant. Because the quantities at the ends of this chain of inequalities are equal, all integrands must be equal, $H = \sqrt{K}$. By our remarks in the last paragraph, this implies that $d\mathbf{e}_3 = H\,d\mathbf{x}$ with H constant, $H\mathbf{x} = \mathbf{e}_3 + H\mathbf{c}$ with $\mathbf{c}$ a constant vector, $|\mathbf{x} - \mathbf{c}| = (1/H)|\mathbf{e}_3| = 1/H$, $\boldsymbol{\Sigma}$ is a sphere.

8.2. Hypersurfaces

We shall extend our study of surfaces to higher dimensions and at the same time motivate some of the things in the next section on Riemannian geometry.

A *hypersurface* is an n-dimensional manifold $\mathbf{M}$ embedded in $\mathbf{E}^{n+1}$. We denote the moving point on $\mathbf{M}$ by $\mathbf{x}$. Our study is local so we pick a definite unit normal $\mathbf{n}$ at each point $\mathbf{x}$ of $\mathbf{M}$. The map $\mathbf{x} \longrightarrow \mathbf{n}$ is a smooth map on $\mathbf{M}$ into $\mathbf{S}^n$. (This can be done globally on a hypersurface $\mathbf{M}$ precisely when $\mathbf{M}$ is orientable.) The tangent space at $\mathbf{x}$ is an n-dimensional Euclidean space; we pick an orthonormal basis for it, $\mathbf{e}_1, \cdots, \mathbf{e}_n$. Thus at $\mathbf{x}$, the vectors $\mathbf{e}_1, \cdots, \mathbf{e}_n, \mathbf{n}$ make up an orthonormal basis of $\mathbf{E}^{n+1}$. Since $d\mathbf{x}$ is in the tangent space we have

$$d\mathbf{x} = \sigma_1\mathbf{e}_1 + \cdots + \sigma_n\mathbf{e}_n$$

where $\sigma_1, \cdots, \sigma_n$ are one-forms on $\mathbf{M}$. From the relations

$$\mathbf{e}_i\cdot\mathbf{e}_k = \delta_{ik}, \qquad \mathbf{e}_i\cdot\mathbf{n} = 0, \qquad \mathbf{n}\cdot\mathbf{n} = 1,$$

we deduce that

$$d\mathbf{e}_i\cdot\mathbf{e}_k + \mathbf{e}_i\cdot d\mathbf{e}_k = 0,$$
$$d\mathbf{e}_i\cdot\mathbf{n} + \mathbf{e}_i\cdot d\mathbf{n} = 0, \qquad \mathbf{n}\cdot d\mathbf{n} = 0$$

and so

$$\begin{cases} d\mathbf{e}_i = \sum \omega_{ij}\mathbf{e}_j - \omega_i\mathbf{n} \\ d\mathbf{n} = \sum \omega_i\mathbf{e}_i \end{cases}$$

where ω_{ij}, ω_i are one-forms on $\mathbf{M}$ and

$$\omega_{ij} + \omega_{ji} = 0.$$

It is convenient to write all of these structure relations in matrix form. We set

$$\begin{cases} \mathbf{e} = \begin{pmatrix} \mathbf{e}_1 \\ \vdots \\ \mathbf{e}_n \end{pmatrix}, & \sigma = (\sigma_1, \cdots, \sigma_n), \\ \Omega = \|\omega_{ij}\|, & \omega = (\omega_1, \cdots, \omega_n). \end{cases}$$

Then

$$\begin{cases} d\mathbf{x} = \sigma\mathbf{e} \\ d\begin{pmatrix} \mathbf{e} \\ \mathbf{n} \end{pmatrix} = \begin{pmatrix} \Omega & -{}^t\omega \\ \omega & 0 \end{pmatrix}\begin{pmatrix} \mathbf{e} \\ \mathbf{n} \end{pmatrix} \\ \Omega + {}^t\Omega = 0. \end{cases}$$

By taking exterior derivatives we obtain integrability conditions. (We shall omit the symbol "$\wedge$" in what follows. All products of differentials are exterior.)

$$\begin{aligned} 0 = d(d\mathbf{x}) &= (d\sigma)\,\mathbf{e} - \sigma(d\mathbf{e}) \\ &= (d\sigma)\,\mathbf{e} - \sigma(\Omega\mathbf{e} - {}^t\omega\mathbf{n}) \\ &= (d\sigma - \sigma\Omega)\,\mathbf{e} + \sigma\,{}^t\omega\mathbf{n}, \end{aligned}$$

$$d\sigma = \sigma\Omega, \qquad \sigma\,{}^t\omega = 0;$$

$$\begin{aligned} 0 &= d\left[d\begin{pmatrix} \mathbf{e} \\ \mathbf{n} \end{pmatrix}\right] \\ &= \begin{pmatrix} d\Omega & -d\,{}^t\omega \\ d\omega & 0 \end{pmatrix}\begin{pmatrix} \mathbf{e} \\ \mathbf{n} \end{pmatrix} - \begin{pmatrix} \Omega & -{}^t\omega \\ \omega & 0 \end{pmatrix} d\begin{pmatrix} \mathbf{e} \\ \mathbf{n} \end{pmatrix} \\ &= \begin{pmatrix} d\Omega & -d\,{}^t\omega \\ d\omega & 0 \end{pmatrix}\begin{pmatrix} \mathbf{e} \\ \mathbf{n} \end{pmatrix} - \begin{pmatrix} \Omega & -{}^t\omega \\ \omega & 0 \end{pmatrix}^2\begin{pmatrix} \mathbf{e} \\ \mathbf{n} \end{pmatrix} \\ &= \begin{pmatrix} d\Omega - \Omega^2 + {}^t\omega\omega & -{}^t(d\omega) + \Omega\,{}^t\omega \\ d\Omega - \omega\Omega & 0 \end{pmatrix}\begin{pmatrix} \mathbf{e} \\ \mathbf{n} \end{pmatrix}, \end{aligned}$$

$$d\Omega - \Omega^2 + {}^t\omega\omega = 0, \qquad d\omega = \omega\Omega.$$

We define a skew-symmetric matrix of two-forms:

$$\Theta = \|\theta_{ij}\| = d\Omega - \Omega^2.$$

We sum up our results:

$$\begin{cases} d\sigma = \sigma\Omega, \\ \Omega + {}^t\Omega = 0, \\ \sigma\,{}^t\omega = 0, \\ d\omega = \omega\Omega, \\ \Theta + {}^t\omega\omega = 0, \end{cases}$$

or in terms of individual elements of the matrices,

$$\begin{cases} d\sigma_j = \sum \sigma_i \omega_{ij}, \\ \omega_{ij} + \omega_{ji} = 0, \\ \sum \sigma_i \omega_i = 0, \\ d\omega_j = \sum \omega_i \omega_{ij}, \\ \theta_{ij} + \omega_i \omega_j = 0. \end{cases}$$

The σ_i form a basis for one-forms on $\mathbf{M}$, hence we have relations

$$\omega_i = \sum b_{ij} \sigma_j .$$

Because $\sum \sigma_i \omega_i = 0$, the b_{ij} must be symmetric,

$$b_{ij} = b_{ji} .$$

The *mean curvature H* and *Gaussian curvature K* are defined by

$$H = \frac{1}{n} \sum b_{ii}, \qquad K = |b_{ij}|$$

Since $\sigma_1 \cdots \sigma_n$ is the n-dimensional volume element on $\mathbf{M}$ and $\omega_1 \cdots \omega_n$ is the corresponding quantity for $\mathbf{S}^n$, K represents the ratio of volumes, volume of spherical image over volume of $\mathbf{M}$, due to

$$\begin{aligned} \omega_1 \cdots \omega_n &= \left(\sum b_{1j} \sigma_j\right) \cdots \left(\sum b_{nj} \sigma_j\right) = |b_{ij}| \sigma_1 \cdots \sigma_n \\ &= K \sigma_1 \cdots \sigma_n . \end{aligned}$$

Suppose one has a function $\mathbf{v} = \mathbf{v}(y, z, \cdots)$ of several variables where $\mathbf{v}$ is always a tangent vector to $\mathbf{M}$. For example, we might assign to each point of a curve on $\mathbf{M}$ a tangent vector at that point, arriving at a vector-valued function of one variable.

How does an observer constrained to $\mathbf{M}$ observe the motion of $\mathbf{v}$? We write

$$\mathbf{v} = \sum c_i \mathbf{e}_i$$

where the c_i are functions and have

$$\begin{aligned} d\mathbf{v} &= \sum dc_i \, \mathbf{e}_i + \sum c_i \, d\mathbf{e}_i \\ &= \sum dc_i \, \mathbf{e}_i + \sum c_i \left(\sum \omega_{ij} \mathbf{e}_j - \omega_i \mathbf{n}\right) \\ &= \sum \left(dc_j + \sum c_i \omega_{ij}\right) \mathbf{e}_j - \left(\sum c_i \omega_i\right) \mathbf{n}. \end{aligned}$$

Our observer who is constrained to move in the hypersurface $\mathbf{M}$ cannot "see" the motion of $\mathbf{v}$ which takes place in the direction normal to $\mathbf{M}$; he sees only the tangential motion of $\mathbf{v}$. Consequently he believes $\mathbf{v}$ is motionless provided

$$\left(dc_j + \sum c_i \omega_{ij}\right) \mathbf{e}_j = 0;$$

that is,

$$dc_j + \sum c_i\omega_{ij} = 0 \qquad (j = 1, \cdots, n).$$

A vector function for which these equations are valid is said to move by *parallel displacement*.

The following can be checked. If $\mathbf{v} = \mathbf{v}(y, z, \cdots)$ and $\mathbf{w} = \mathbf{w}(y, z, \cdots)$ are two such vector-valued functions which are compatible [for each point $(y, z, \cdots)$ in the parameter space $\mathbf{v}$ and $\mathbf{w}$ are tangent at the same point of $\mathbf{M}$] and each moves by parallel displacement, then $\mathbf{v}\cdot\mathbf{w}$ is constant. In particular, $|\mathbf{v}|^2 = \mathbf{v}\cdot\mathbf{v}$ is constant.

Let $P = P(s)$ be a curve on $\mathbf{M}$ parametrized by its arc length s so that

$$\mathbf{t} = t(s) = \frac{dP}{ds}$$

is the unit tangent vector. The curve is called a *geodesic* provided $\mathbf{t}$ moves by parallel displacement.

There is a geometric interpretation of the matrix $\|b_{ij}\|$ which is quite fundamental. To each particular displacement $d\mathbf{x}$ of the position vector $\mathbf{x}$ corresponds a displacement $d\mathbf{n}$ of the unit normal $\mathbf{n}$. Both $d\mathbf{x}$ and $d\mathbf{n}$ are in the tangent space at $\mathbf{x}$ so that we can look on

$$d\mathbf{x} \longrightarrow d\mathbf{n}$$

as a linear transformation A of this tangent space. More precisely, let $\mathbf{v}$ be any tangent vector at $\mathbf{x}$. Pick any curve $\mathbf{x} = \mathbf{x}(t)$ through $\mathbf{x}$ so that

$$\left.\frac{d\mathbf{x}}{dt}\right|_0 = \mathbf{v}.$$

We follow the normal $\mathbf{n} = \mathbf{n}(t)$ as $\mathbf{x}$ traverses the curve. Then

$$\left.\frac{d\mathbf{n}}{dt}\right|_0 = A\mathbf{v}$$

is our definition of A. We see that this is quite independent of the choice of the curve $\mathbf{x}(t)$ so long as it has the prescribed tangent $\mathbf{v}$ at $t = 0$.

For suppose

$$\mathbf{v} = c_1\mathbf{e}_1 + \cdots + c_n\mathbf{e}_n.$$

Then

$$d\mathbf{x} = \sum \sigma_i \mathbf{e}_i$$

so that

$$\left.\frac{\sigma_i}{dt}\right|_0 = c_i.$$

But

$$\left.\frac{d\mathbf{n}}{dt}\right|_0 = \sum \left.\frac{\omega_i}{dt}\right|_0 \mathbf{e}_i.$$

Now

$$\omega_i = \sum b_{ij}\sigma_j,$$

$$\left.\frac{\omega_i}{dt}\right|_0 = \sum b_{ij} \left.\frac{\sigma_j}{dt}\right|_0 = \sum b_{ij}c_j$$

so our result is

$$A(\sum c_i \mathbf{e}_i) = \sum(\sum b_{ij}c_j)\,\mathbf{e}_i$$

which establishes these points: (1) A is a well-defined function on the tangent space at $\mathbf{x}$ to itself, (2) A is linear, (3) the matrix representation of A with respect to the basis $\mathbf{e}$ is $\|b_{ij}\|$.

Since the matrix $\|b_{ij}\|$ is symmetric, the linear transformation A on the (Euclidean) n-dimensional tangent space at $\mathbf{x}$ is *self-adjoint*: for each pair of tangent vectors $\mathbf{v}$, $\mathbf{w}$,

$$(A\mathbf{v})\cdot\mathbf{w} = \mathbf{v}\cdot(A\mathbf{w}).$$

It is clear that our definition of A depends only on the hypersurface $\mathbf{M}$ and the way it is embedded in $\mathbf{E}^{n+1}$, not on our choice of the moving frame $\mathbf{e}$. Consequently the formulas

$$H = \frac{1}{n}\operatorname{trace}(A)$$

$$K = |A|$$

show that H and K are geometric quantities.

Since A is self-adjoint, its characteristic roots are all real and they are called *principal curvatures*. The corresponding characteristic vectors define (in general) n direction fields on $\mathbf{M}$ called *principal directions* whose integral curves are *lines of curvature*.

Convexity of $\mathbf{M}$ can be interpreted in terms of a definiteness condition on the transformation A. To make this precise, suppose $\mathbf{M}$ is convex and we choose for $\mathbf{n}$ the inward unit normal.

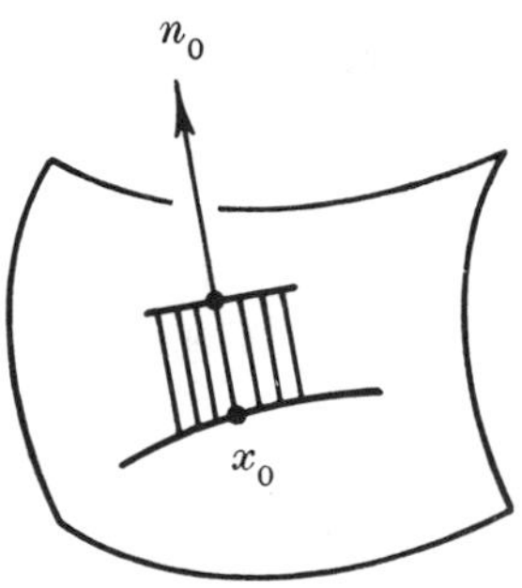

We fix a point $\mathbf{x}_0$ on $\mathbf{M}$ and form a corresponding *normal section*, the curve of intersection of $\mathbf{M}$ with any (two-dimensional) plane on the line $\mathbf{n}_0$. This is a curve $\mathbf{x} = \mathbf{x}(t)$ which is convex on its plane, which means that the function

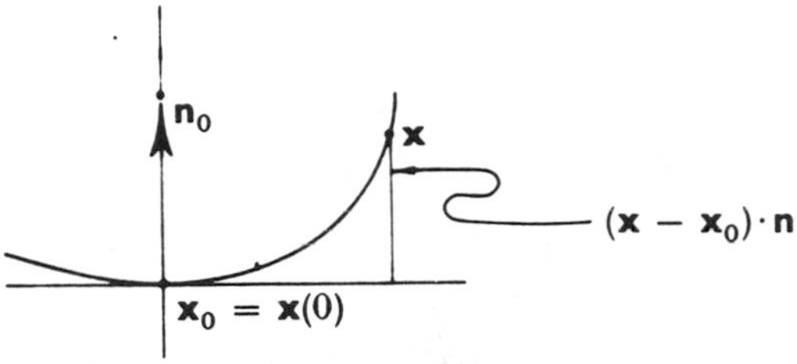

$$f(t) = (\mathbf{x} - \mathbf{x}_0)\cdot\mathbf{n}_0$$

is convex, hence satisfies

$$\left.\frac{d^2f}{dt^2}\right|_{t=0} \geqq 0.$$

We have

$$\frac{df}{dt} = \left(\frac{d\mathbf{x}(t)}{dt}\right)\cdot\mathbf{n}_0 = \left(\sum \frac{\sigma_i}{dt}\,\mathbf{e}_i\right)\cdot\mathbf{n}_0,$$

$$\frac{d^2f}{dt^2} = \sum \frac{d}{dt}\left(\frac{\sigma_i}{dt}\right)\mathbf{e}_i\cdot\mathbf{n}_0 + \sum\left(\frac{\sigma_i}{dt}\right)\left(\frac{d\mathbf{e}_i}{dt}\right)\cdot\mathbf{n}_0.$$

Now

$$\frac{d\mathbf{e}_i}{dt}\cdot\mathbf{n}_0 = \sum\left(\frac{\omega_{ij}}{dt}\right)\mathbf{e}_j\cdot\mathbf{n}_0 - \left(\frac{\omega_i}{dt}\right)\mathbf{n}\cdot\mathbf{n}_0.$$

Since the tangent vectors $\mathbf{e}_i(0)$ at $\mathbf{x}(0) = \mathbf{x}_0$ are orthogonal to the normal $\mathbf{n}(0) = \mathbf{n}_0$, the condition reduces to

$$\sum \left.\left(\frac{\sigma_i}{dt}\right)\right|_0 \left.\left(\frac{\omega_i}{dt}\right)\right|_0 \leqq 0.$$

But $\omega_i = \sum b_{ij}\sigma_j$, so we have

$$\sum b_{ij}\left.\left(\frac{\sigma_i}{dt}\right)\right|_0 \left.\left(\frac{\sigma_j}{dt}\right)\right|_0 \leqq 0.$$

Since the direction

$$\left.\left(\frac{\sigma_1}{dt}, \ldots, \frac{\sigma_n}{dt}\right)\right|_0$$

is arbitrary, we conclude from this that the matrix

$$\|b_{ij}\|$$

is negative semidefinite so that the same is true for the transformation A which this symmetric matrix represents.

(The correctness of sign can be verified in the simplest possible case, that of a convex curve in the plane. The usual Frenet formulas are

$$\begin{cases} d\mathbf{x} = ds\,\mathbf{e}_1 \\ d\mathbf{e}_i = \kappa\, ds\,\mathbf{n} \\ d\mathbf{n} = -\kappa\, ds\,\mathbf{e}_1 \qquad (s = \text{arc length}) \end{cases}$$

so that $\sigma_1 = ds$, $\omega_1 = -\kappa\, ds$, $b_{11} = -\kappa \leqq 0$.)

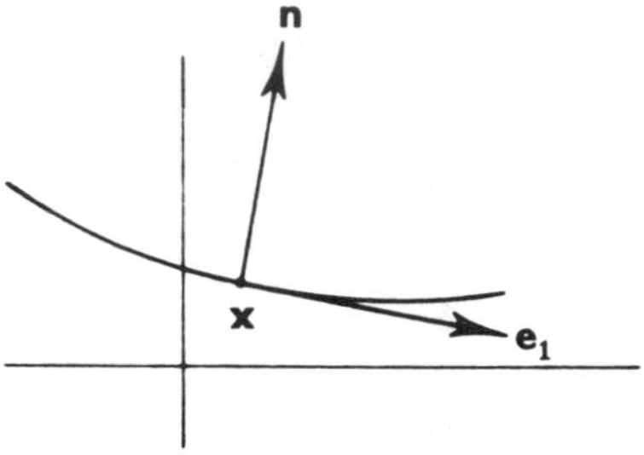

We return to the general situation. The elements θ_{ij} of the *curvature matrix* Θ are *curvature forms*. We may write

$$\theta_{ij} = \tfrac{1}{2} \sum R_{ijkl}\sigma_k\sigma_l, \qquad R_{ijkl} + R_{ijlk} = 0,$$

defining the *Riemann curvature tensor* R_{ijkl} of the hypersurface. Because of the relations

$$\theta_{ij} + \omega_i\omega_j = 0,$$

and

$$\omega_i\omega_j = \sum b_{ik}b_{jl}\sigma_k\sigma_l = \tfrac{1}{2} \sum (b_{ik}b_{jl} - b_{il}b_{jk})\,\sigma_k\sigma_l$$

we have

$$R_{ijkl} + \begin{vmatrix} b_{ik} & b_{il} \\ b_{jk} & b_{jl} \end{vmatrix} = 0.$$

Algebraic consequences of these formulas are the following:

$$\begin{cases} R_{ijkl} + R_{ijlk} = 0, \\ R_{ijkl} + R_{jikl} = 0, \\ R_{ijkl} + R_{iklj} + R_{iljk} = 0, \\ R_{ijkl} = R_{klij}, \end{cases}$$

all of which follow easily.

We shall see that the Riemann tensor is independent of how $\mathbf{M}$ is embedded in $\mathbf{E}^{n+1}$ so that these relations are particularly interesting, connecting the intrinsic Riemann tensor with the quantities b_{ij}, which clearly depend on the embedding.

Indeed, it turns out that the R's are determined by the σ's alone with no reference to the normal $\mathbf{n}$. This means that if two hypersurfaces $\mathbf{M}_1$ and $\mathbf{M}_2$ are in a one-one correspondence which preserves distance (i.e., preserves the Euclidean geometries in each pair of corresponding tangent hyperplanes), then $\mathbf{M}_1$ and $\mathbf{M}_2$ have the same Riemann curvature tensor.

What we shall show is that the equations

$$d\sigma = \sigma\Omega, \qquad \Omega + {}^t\Omega = 0$$

determine Ω uniquely, so that Ω is completely determined by the σ's alone. This is more in context in general Riemannian geometry so we shall postpone the proof until p. 129 of the next section. Having this result, it follows that

$$\Theta = d\Omega - \Omega^2$$

is also completely determined by the σ's, hence so is the Riemann tensor.

We shall now take up the special case in which our hypersurface $\mathbf{M}$ is given in the form

$$u = u(x^1, \cdots, x^n)$$

in $x^1, \cdots, x^n$, u-space.

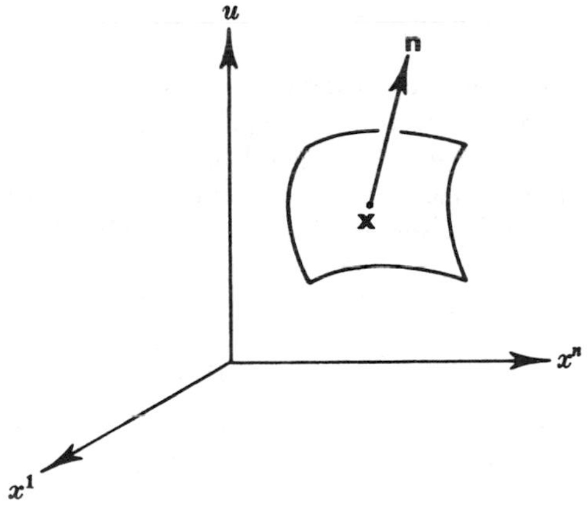

It is convenient to set

$$p_i = \frac{\partial u}{\partial x^i}, \qquad r_{ij} = \frac{\partial^2 u}{\partial x^i \partial x^j}.$$

We have for the position vector

$$\mathbf{x} = (x^1, x^2, \cdots, x^n, u)$$

and so

$$\begin{aligned} d\mathbf{x} &= (dx^1, dx^2, \cdots, dx^n, du) \\ &= (dx^1, \cdots, dx^n, \sum p_i\, dx^i) \\ &= \sum \mathbf{t}_i\, dx^i \end{aligned}$$

where

$$\mathbf{t}_i = (\delta_{i1}, \cdots, \delta_{in}, p_i).$$

The vectors $\mathbf{t}_i, \cdots, \mathbf{t}_n$ are tangent vectors and evidently form a linear basis, but generally *not* an orthonormal basis, of the tangent hyperplane.

The vector

$$\mathbf{w} = (-p_1, \cdots, -p_n, 1)$$

satisfies

$$\mathbf{w} \cdot \mathbf{t}_i = 0$$

so it is a normal vector (with positive component in the u-direction). The unit normal $\mathbf{n}$ is given by

$$\mathbf{w} = w\mathbf{n}$$

where

$$w^2 = \mathbf{w} \cdot \mathbf{w} = 1 + \sum p_i^{\,2}.$$

We note that

$$w\,dw = \sum p_i\, dp_i = \sum p_i r_{ij}\, dx^j.$$

We shall now determine the matrix representation of the basic linear transformation A with respect to the basis $\mathbf{t}_1, \cdots, \mathbf{t}_n$ of the tangent hyperplane. This matrix is not symmetric in general because $(\mathbf{t}_i)$ is not an orthonormal basis. However, its trace and determinant are the trace and determinant of A since <u>any</u> matrix representation yields a valid determination of these quantities.

Suppose the matrix we seek is $\|a_{ij}\|$. Then this means

$$A\mathbf{t}_i = \sum a_{ij}\mathbf{t}_j.$$

Let $\mathbf{v}$ be *any* tangent vector. Because of these relations plus the fact that A is self-adjoint we have

$$(A\mathbf{v}) \cdot \mathbf{t}_i = (A\mathbf{t}_i) \cdot \mathbf{v} = \sum a_{ij}(\mathbf{t}_j \cdot \mathbf{v}).$$

Now going back to the very definition of A, symbolically,

$$A: \quad d\mathbf{x} \longrightarrow d\mathbf{n},$$

we see that this relation means that

$$(d\mathbf{n}) \cdot \mathbf{t}_i = \sum a_{ij}(d\mathbf{x} \cdot \mathbf{t}_j).$$

It is from this set of equations that we shall determine $\|a_{ij}\|$. We have

$$d\mathbf{w} = dw\,\mathbf{n} + w\,d\mathbf{n},$$

$$d\mathbf{w}\cdot\mathbf{t}_i = w(d\mathbf{n})\cdot\mathbf{t}_i$$

since $\mathbf{n}\cdot\mathbf{t}_i = 0$. But

$$\begin{aligned} d\mathbf{w}\cdot\mathbf{t}_i &= (-dp_1, \cdots, -dp_n, 0)\cdot(\delta_{i1}, \cdots, \delta_{in}, p_i) \\ &= -dp_i = -\sum r_{ij}\,dx^j \end{aligned}$$

so we have obtained

$$d\mathbf{n}\cdot\mathbf{t}_i = -\frac{1}{w}\sum r_{ij}\,dx^j.$$

On the other hand,

$$\begin{aligned} d\mathbf{x}\cdot\mathbf{t}_j &= (dx^1, \cdots, dx^n, du)\cdot(\delta_{j1}, \cdots, \delta_{jn}, p_j) \\ &= dx^j + p_j\,du = dx^j + p_j\sum p_k\,dx^k \\ &= \sum_k (\delta_{jk} + p_j p_k)\,dx^k, \end{aligned}$$

so we have

$$\sum a_{ij}(\delta_{jk} + p_j p_k)\,dx^k = -\frac{1}{w}\sum r_{ik}\,dx^k,$$

$$\sum a_{ij}(\delta_{jk} + p_j p_k) = -\frac{1}{w}\,r_{ik}.$$

Setting

$$R = \|r_{ik}\|, \qquad \mathbf{p} = \begin{pmatrix} p_1 \\ \vdots \\ p_n \end{pmatrix}, \qquad A = \|a_{ij}\|,$$

we may write this as

$$A(I + \mathbf{p}\,{}^t\mathbf{p}) = -\frac{1}{w}\,R, \qquad A = -\frac{1}{w}\,R(I + \mathbf{p}\,{}^t\mathbf{p})^{-1}.$$

Since

$$\mathbf{p}\,{}^t\mathbf{p} = \sum p_i^2 = w^2 - 1$$

we see that

$$(I + \mathbf{p}\,{}^t\mathbf{p})\left(I - \left(\frac{1}{w^2}\right)\mathbf{p}\,{}^t\mathbf{p}\right) = I + \mathbf{p}\,{}^t\mathbf{p} - \left(\frac{1}{w^2}\right)\mathbf{p}\,{}^t\mathbf{p} - \frac{w^2 - 1}{w^2}\,\mathbf{p}\,{}^t\mathbf{p} = I,$$

hence

$$(I + \mathbf{p}\,{}^t\mathbf{p})^{-1} = I - \left(\frac{1}{w^2}\right)\mathbf{p}\,{}^t\mathbf{p},$$

$$A = -\frac{1}{w}\,R\left(I - \left(\frac{1}{w^2}\right)\mathbf{p}\,{}^t\mathbf{p}\right).$$

The mean curvature H is found by taking the trace:

$$H = \frac{1}{n}\operatorname{tr}(A) = -\frac{1}{nw}\operatorname{tr}\left[R - \frac{1}{w^2}R\mathbf{p}\,{}^t\mathbf{p}\right]$$
$$= -\frac{1}{nw}\left[\sum r_{ii} - \sum\left(\frac{1}{w^2}\right)p_i r_{ij} p_j\right]$$
$$= -\frac{1}{nw}\left[\Delta u - \sum \frac{1}{w^2} p_i r_{ij} p_j\right]$$

where Δu is the Laplacian of u.

The Gaussian curvature K is found by taking the determinant:

$$K = |A| = \frac{(-1)^n}{w^n}|R||I + \mathbf{p}\,{}^t\mathbf{p}|^{-1}.$$

One finds by a short calculation† that

$$|I + \mathbf{p}\,{}^t\mathbf{p}| = w^2$$

so that

$$K = \frac{(-1)^n}{w^{n+2}}|R|.$$

For the special case $n = 2$ of surfaces, we use the standard Monge notation

$$p = \frac{\partial u}{\partial x}, \qquad q = \frac{\partial u}{\partial y}, \qquad r = \frac{\partial^2 u}{\partial x^2}, \qquad s = \frac{\partial^2 u}{\partial x\,\partial y}, \qquad t = \frac{\partial^2 u}{\partial y^2}$$

and have

$$w^2 = 1 + p^2 + q^2,$$

$$H = \frac{-1}{2w^3}[w^2(r + t) - (p^2 r + 2pqs + q^2 t)]$$
$$= \frac{-1}{2w^3}[(1 + p^2 + q^2)(r + t) - (p^2 r + 2pqs + q^2 t)]$$
$$= \frac{-1}{2w^3}[(1 + q^2)r - 2pqs + (1 + p^2)t],$$

and

$$K = \frac{(rt - s^2)}{w^4},$$

both familiar formulas.

† Set $\mathbf{e}_1 = (1, 0, \ldots, 0)$, $\mathbf{e}_2 = (0, 1, \ldots, 0), \ldots, \mathbf{e}_n = (0, 0, \ldots, 0, 1)$. Then

$$\begin{aligned}|I + \mathbf{p}\,{}^t\mathbf{p}| &= |\mathbf{e}_1 + p_1\,{}^t\mathbf{p}, \ldots, \mathbf{e}_n + p_n\,{}^t\mathbf{p}| \\ &= |\mathbf{e}_1, \ldots, \mathbf{e}_n| + \sum p_i \cdot |\mathbf{e}_1, \ldots, \mathbf{e}_{i-1}, {}^t\mathbf{p}, \mathbf{e}_{i+1}, \ldots, \mathbf{e}_n| \\ &= 1 + \sum p_i^2 = w^2.\end{aligned}$$

8.3. Riemannian Geometry, Local Theory

The problem here is to deal with the inner geometry of a manifold which is not part of a Euclidean space. If the manifold were part of Euclidean space, it would inherit a local Euclidean geometry (distance function) from that of the including space, as was the case for the hypersurfaces discussed in the last section. However, it is not part of Euclidean space, so we must postulate the existence of a local distance geometry. What we do in effect is to presuppose that each tangent space possesses an inner product which is smooth.

Thus we let $\mathbf{M}$ be an n-dimensional manifold. We suppose that an inner product is given in the tangent space at each point P of $\mathbf{M}$. Thus if $\mathbf{v}$ and $\mathbf{w}$ are two tangent vectors at the same point P, $\mathbf{v}\cdot\mathbf{w}$ is a real number. The inner product is supposed to be smooth in this sense: If $\mathbf{v}$ and $\mathbf{w}$ are vector fields on $\mathbf{M}$, then $\mathbf{v}\cdot\mathbf{w}$ is a smooth function on $\mathbf{M}$. (Precisely, at each point P of $\mathbf{M}$, the values of the given fields $\mathbf{v}$, $\mathbf{w}$ at P are $\mathbf{v}_P$, $\mathbf{w}_P$, tangent vectors at P, and we are requiring that $\mathbf{v}_P\cdot\mathbf{w}_P$ be a smooth function of P.)

The procedure in Section 2.5 (pp. 13–14) for finding an orthonormal basis may be made constructive, smooth operations at each step. We know that there exists on each local coordinate neighborhood on $\mathbf{M}$ a set of n vector fields, forming a basis for the tangent space at each point of the neighborhood. We convert these fields to orthonormal ones to arrive at smooth vector fields

$$\mathbf{e}_1, \cdots, \mathbf{e}_n$$

defined on the local coordinate neighborhood in question and satisfying

$$\mathbf{e}_i\cdot\mathbf{e}_j = \delta_{ij}.$$

Pretending for a moment that we are observers constrained to the manifold $\mathbf{M}$, we set out on a somewhat symbolic voyage, hoping in doing so to motivate the right steps. We let P denote the moving point on $\mathbf{M}$ and wish to think of its arbitrary displacement dP as a tangent vector with differential form coefficients; we hopefully write

$$\mathbf{d}P = \sum \sigma_i \mathbf{e}_i$$

with $\sigma_1, \cdots, \sigma_n$ differential one-forms on our neighborhood. (Since $\mathbf{d}P$ is in no sense an exterior derivative of anything, we distinguish this $\mathbf{d}$ by bold-face type from the usual d. The same applies to the $\mathbf{de}_i$ below.) We expect to arrive at a basis $\sigma_1, \cdots, \sigma_n$ for the one-forms which in some sense is orthonormal, dual to the basis $\mathbf{e}_1, \cdots, \mathbf{e}_n$ of vectors.

We must be guided by our experience in Euclidean space. There we would take a coordinate system $u^1, \cdots, u^n$ and without hesitation write

$$\mathbf{d}P = \sum du^i\left(\frac{\partial}{\partial u^i}\right),$$

the vectors

$$\frac{\partial}{\partial u^1}, \cdots, \frac{\partial}{\partial u^n}$$

being the natural frame associated with the coordinate system. But precisely this can be done on $\mathbf{M}$. The expression

$$\mathbf{d}P = \sum du^i\left(\frac{\partial}{\partial u^i}\right)$$

is independent of the local coordinate system. Indeed, if $\bar{u}^1, \cdots, \bar{u}^n$ is another system, then

$$d\bar{u}^i = \sum \frac{\partial \bar{u}^i}{\partial u^j} du^j,$$

$$\left(\frac{\partial}{\partial \bar{u}^i}\right) = \sum \frac{\partial u^k}{\partial \bar{u}^i}\left(\frac{\partial}{\partial u^k}\right),$$

and so

$$\sum d\bar{u}^i\left(\frac{\partial}{\partial \bar{u}^i}\right) = \sum \frac{\partial \bar{u}^i}{\partial u^j}\frac{\partial u^k}{\partial \bar{u}^i} du^j\left(\frac{\partial}{\partial u^k}\right)$$

$$= \sum \delta_j{}^k du^j\left(\frac{\partial}{\partial u^k}\right) = \sum du^j\left(\frac{\partial}{\partial u^j}\right);$$

that is, $\mathbf{d}P$ is the same either way.

Having this, we express the natural frame in terms of the orthonormal one,

$$\left(\frac{\partial}{\partial u^i}\right) = \sum a_{ij}\,\mathbf{e}_j$$

and solve for the σ_j:

$$\sum du^i\left(\frac{\partial}{\partial u^i}\right) = \mathbf{d}P = \sum \sigma_j\,\mathbf{e}_j,$$

$$\sum du^i a_{ij}\,\mathbf{e}_j = \sum \sigma_j\,\mathbf{e}_j,$$

$$\sigma_j = \sum a_{ij}\,du^i.$$

We have reached our first equation of structure for local Riemannian geometry:

$$\mathbf{d}P = \sum \sigma_i\,\mathbf{e}_i.$$

Our next venture is to attempt an analogue to the equations for the displacements $\mathbf{de}_i$ of the vectors of the moving frame. Here we make an essential departure from what we did with surfaces and hypersurfaces; we must restrict our attention to "the tangential component" of $\mathbf{de}_i$ for the simple reason that we are constrained to $\mathbf{M}$ and can see no happenings in any "normal" direction. Thus we seek expressions

$$\mathbf{de}_i = \sum \omega_{ij}\,\mathbf{e}_j$$

with one-forms ω_{ij}. We try to find such ω_{ij} so as to be consistent with these conditions:

$$\mathbf{de}_i \cdot \mathbf{e}_k + \mathbf{e}_i \cdot \mathbf{de}_k = 0, \tag{1}$$

$$\mathbf{d}(\mathbf{d}P) = 0. \tag{2}$$

The reason for (1) is that $d(\mathbf{e}_i \cdot \mathbf{e}_k) = d(\delta_{ik}) = 0$ and we hope to "differentiate" the dot product by the usual product rule. In choosing (2) we simply go according to the Euclidean analogue. In the usual way, (1) reduces to

$$\omega_{ik} + \omega_{ki} = 0. \tag{1'}$$

We explore condition (2):

$$\mathbf{d}(\textstyle\sum \sigma_i \mathbf{e}_i) = 0,$$
$$\textstyle\sum d\sigma_i \mathbf{e}_i - \sum \sigma_i \mathbf{de}_i = 0,$$
$$\textstyle\sum (d\sigma_j - \sum \sigma_i \omega_{ij}) \mathbf{e}_j = 0,$$

so (2) is equivalent to

$$d\sigma_j = \textstyle\sum \sigma_i \omega_{ij}. \tag{2'}$$

We have finally come to a well-formulated problem: given the basis $\sigma_1, \cdots, \sigma_n$ of one-forms, find one-forms ω_{ij} satisfying

$$\begin{cases} (1') & \omega_{ij} + \omega_{ji} = 0 \\ (2') & d\sigma_i = \sum \sigma_j \omega_{ji}. \end{cases}$$

We shall show that *this problem has exactly one solution.* (This completes a point we left open on p. 123 of the last section.)

Since the σ_i form a basis we may write

$$\omega_{ij} = \textstyle\sum \Gamma_{ijk} \sigma_k$$

where the (unknown) functions Γ_{ijk} are the *connection coefficients*, or *Christoffel symbols*. Equivalent to (1′) is

$$\Gamma_{ijk} + \Gamma_{jik} = 0. \tag{1''}$$

The $d\sigma_i$ are known since the σ_i are known; we may write

$$d\sigma_i = \tfrac{1}{2} \textstyle\sum c_{ijk} \sigma_j \sigma_k, \qquad c_{ijk} + c_{ikj} = 0.$$

We have

$$\begin{aligned} d\sigma_i &= \tfrac{1}{2} \textstyle\sum c_{ijk} \sigma_j \sigma_k = \sum \sigma_j \omega_{ji} \\ &= \textstyle\sum \sigma_j \sum \Gamma_{jik} \sigma_k = \tfrac{1}{2} \sum (\Gamma_{jik} - \Gamma_{kij}) \sigma_j \sigma_k, \end{aligned}$$

and so (2′) is equivalent to

$$\Gamma_{jik} - \Gamma_{kij} = c_{ijk}. \tag{2''}$$

Our precise statement now is this:

Given functions c_{ijk} *such that* $c_{ijk} + c_{ikj} = 0$, *the system of linear equations*

$$\begin{cases} \Gamma_{ijk} + \Gamma_{jik} = 0 \\ \Gamma_{jik} - \Gamma_{kij} = c_{ijk} \end{cases}$$

has a unique solution given by

$$\Gamma_{ijk} = \tfrac{1}{2}(c_{kij} - c_{jki} - c_{ijk}).$$

For if Γ_{ijk} is any solution, then we use the equations alternately to derive

$$\begin{aligned} \Gamma_{ijk} &= -\Gamma_{jik} = -\Gamma_{kij} - c_{ijk} \\ &= \Gamma_{ikj} - c_{ijk} = \Gamma_{jki} + c_{kij} - c_{ijk} \\ &= -\Gamma_{kji} + c_{kij} - c_{ijk} = -\Gamma_{ijk} - c_{jki} + c_{kij} - c_{ijk}, \end{aligned}$$

hence

$$2\Gamma_{ijk} = c_{kij} - c_{jki} - c_{ijk}$$

which establishes the uniqueness of solution. It is easily verified that the asserted values of Γ_{ijk} really are a solution.

We have completed our structure equations,

$$\begin{cases} \mathbf{d}P = \boldsymbol{\sigma}\mathbf{e} \\ \mathbf{de} = \Omega\mathbf{e} \\ \Omega + {}^t\Omega = 0, \end{cases}$$

where we have introduced the matrix notation

$$\boldsymbol{\sigma} = (\sigma_1, \cdots, \sigma_n), \qquad \mathbf{e} = \begin{pmatrix} \mathbf{e}_1 \\ \vdots \\ \mathbf{e}_n \end{pmatrix}, \qquad \Omega = \|\omega_{ij}\|$$

and already have one integrability condition,

$$d\boldsymbol{\sigma} = \boldsymbol{\sigma}\Omega.$$

There is no reason for believing that $\mathbf{d}(\mathbf{de}) = 0$ in any sense. We have

$$\begin{aligned} \mathbf{d}^2\mathbf{e} = \mathbf{d}(\mathbf{de}) = \mathbf{d}(\Omega\mathbf{e}) &= (d\Omega)\,\mathbf{e} - \Omega(\mathbf{de}) \\ &= (d\Omega - \Omega^2)\,\mathbf{e}. \end{aligned}$$

We set

$$\Theta = \|\theta_{ij}\| = d\Omega - \Omega^2,$$

the *curvature matrix* which appears from the symbolic equation

$$\mathbf{d}^2\mathbf{e} = \Theta\mathbf{e}$$

as representing a "second derivative"—exactly how one thinks of curvature in elementary differential geometry.

We derive further integrability conditions by differentiating. From

$$d\boldsymbol{\sigma} = \boldsymbol{\sigma}\Omega$$

we have

$$\begin{aligned} 0 = d(d\boldsymbol{\sigma}) &= (d\boldsymbol{\sigma})\,\Omega - \boldsymbol{\sigma}(d\Omega) \\ &= (\boldsymbol{\sigma}\Omega)\,\Omega - \boldsymbol{\sigma}(d\Omega), \end{aligned}$$

hence

$$\boldsymbol{\sigma}\Theta = 0.$$

From

$$\Theta = d\Omega - \Omega^2$$

we have

$$\begin{aligned} d\Theta &= d(d\Omega) - d(\Omega^2) \\ &= 0 - (d\Omega)\,\Omega + \Omega(d\Omega) \\ &= -(\Theta + \Omega^2)\,\Omega + \Omega(\Theta + \Omega^2), \end{aligned}$$

hence

$$d\Theta = \Omega\Theta - \Theta\Omega,$$

which comprises the *Bianchi identity*.

We sum up:

Structure equations	Integrability conditions
$\left\{\begin{aligned} \mathbf{d}P &= \boldsymbol{\sigma}\mathbf{e} \\ \mathbf{de} &= \Omega\mathbf{e} \\ \Omega + {}^t\Omega &= 0 \\ \Theta &= d\Omega - \Omega^2 \end{aligned}\right\}$	$\left\{\begin{aligned} d\boldsymbol{\sigma} &= \boldsymbol{\sigma}\Omega \\ \boldsymbol{\sigma}\Theta &= 0 \\ d\Theta &= \Omega\Theta - \Theta\Omega. \end{aligned}\right\}$

The n form $\sigma_1 \cdots \sigma_n$ is the *volume element* of $\mathbf{M}$. It is determined up to sign. If $\mathbf{M}$ is oriented, one may fix it by choosing only moving frames coherent to the orientation.

The θ_{ij} are two-forms which may be written

$$\theta_{ij} = \tfrac{1}{2} \sum R_{ijkl}\sigma_k\sigma_l$$

which defines the *Riemann curvature tensor*. We have

$$\begin{aligned} R_{ijkl} + R_{ijlk} &= 0 \\ R_{ijkl} + R_{jikl} &= 0. \end{aligned}$$

The relation $\boldsymbol{\sigma}\Theta = 0$, or

$$\sum R_{ijkl}\sigma_i\sigma_k\sigma_l = 0,$$

is equivalent to

$$R_{ijkl} + R_{iklj} + R_{iljk} = 0.$$

In the special case of a hypersurface we had the symmetry condition

$$R_{ijkl} = R_{klij}$$

as an obvious consequence of the expression for R_{ijkl} as a two-round minor (see p. 122). Such a determinant representation is not possible in this general situation, but it turns out that this symmetry of the Riemann tensor is true anyway, an algebraic consequence of the other relations:

$$\begin{aligned}(R_{ijkl} - R_{klij}) &= R_{ijkl} + (R_{kijl} + R_{kjli}) \\ &= (R_{ijkl} + R_{iklj}) - R_{jkli} \\ &= -R_{iljk} - R_{jkli} \\ &= R_{lijk} + (R_{jlik} + R_{jikl}) \\ &= (R_{lijk} + R_{ljki}) + R_{jikl} \\ &= -R_{lkij} + R_{jikl} \\ &= -(R_{ijkl} - R_{klij}),\end{aligned}$$

and so

$$2(R_{ijkl} - R_{klij}) = 0$$

which implies the symmetry in question.

Those who have been through the mill in Riemannian geometry *à la* classical tensors will be anxious to see the connection. There one deals with a natural frame

$$\mathbf{v}_1 = \frac{\partial}{\partial u^1}, \cdots, \qquad \mathbf{v}_n = \frac{\partial}{\partial u^n}$$

due to a local coordinate system $(u^1, \cdots, u^n)$. One sets

$$g_{ij} = \mathbf{v}_i \cdot \mathbf{v}_j$$

defining the positive definite symmetric matrix (metric tensor)

$$G = \|g_{ij}\|.$$

Usually one looks instead at the corresponding definite quadratic form

$$ds^2 = \sum g_{ij}[du^i du^j]$$

where the brackets remind us that this is ordinary, not exterior, multiplication of differentials. This is motivated by the formula

$$s = \int \left[\sum g_{ij} \left(\frac{du^i}{dt} \right) \left(\frac{du^j}{dt} \right) \right]^{1/2} dt$$

for the arc length of a curve $u^i = u^i(t)$.

Now one has

$$\mathbf{d}P = \sum du^i \, \mathbf{v}_i .$$

We try

$$\mathbf{dv}_i = \sum \eta_i{}^j \mathbf{v}_j$$

and then introduce Christoffel symbols by

$$\eta_i{}^j = \sum \begin{Bmatrix} j \\ ik \end{Bmatrix} du^k.$$

To have

$$dg_{ij} = \mathbf{dv}_i \cdot \mathbf{v}_j + \mathbf{v}_i \cdot \mathbf{dv}_j$$

and

$$0 = \mathbf{d}(\mathbf{d}P) = -\sum du^i \mathbf{dv}_i$$

requires

$$\frac{\partial g_{ij}}{\partial u^k} = \sum \begin{Bmatrix} l \\ ik \end{Bmatrix} g_{lj} + \sum \begin{Bmatrix} l \\ jk \end{Bmatrix} g_{il}$$

$$\begin{Bmatrix} j \\ ik \end{Bmatrix} = \begin{Bmatrix} j \\ ki \end{Bmatrix}.$$

Lowering indices:

$$[i, jk] = \sum \begin{Bmatrix} l \\ jk \end{Bmatrix} g_{il},$$

the equations become

$$[j, ik] + [i, jk] = \frac{\partial g_{ij}}{\partial u^k}$$

$$[j, ik] = [j, ki]$$

with the usual solution

$$[i, jk] = \frac{1}{2}\left(\frac{\partial g_{ij}}{\partial u^k} + \frac{\partial g_{ik}}{\partial u^j} - \frac{\partial g_{jk}}{\partial u^i}\right),$$

and so it goes.

This digression out of the way, we briefly consider parallel displacement. Let $\mathbf{v}$ be a tangent vector on $\mathbf{M}$ which is a function of one or more variables. We write

$$\mathbf{v} = \sum c_i \mathbf{e}_i$$

where the c_i are functions and have

$$\begin{aligned} \mathbf{dv} &= \sum dc_i \mathbf{e}_i + \sum c_i \omega_{ij} \mathbf{e}_j \\ &= \sum (dc_j + \sum c_i \omega_{ij}) \mathbf{e}_j. \end{aligned}$$

A vector $\mathbf{v}$ moves by parallel displacement if $d\mathbf{v} = 0$, i.e.,

$$dc_j + \sum c_i \omega_{ij} = 0.$$

With this interpretation of $\mathbf{dv}$ one has for two compatible vector functions $\mathbf{v}$, $\mathbf{w}$ that

$$d(\mathbf{v} \cdot \mathbf{w}) = (\mathbf{dv}) \cdot \mathbf{w} + \mathbf{v} \cdot (\mathbf{dw}).$$

Consequently if both $\mathbf{v}$ and $\mathbf{w}$ move by parallel displacement then $\mathbf{v} \cdot \mathbf{w}$ is constant; its differential vanishes.

A curve $P = P(s)$ on $\mathbf{M}$ where s is the arc length is called a *geodesic* if the unit tangent $\mathbf{d}P/ds$ moves by parallel displacement.

Example. We consider the upper half plane with the (Poincaré) metric

$$ds^2 = \frac{[dx]^2 + [dy]^2}{y^2}.$$

Here

$$\sigma_1 = \frac{dx}{y}, \qquad \sigma_2 = \frac{dy}{y}.$$

Since the position vector is $P = (x, y)$ we have

$$\mathbf{d}P = (dx, dy) = \sigma_1 \mathbf{e}_1 + \sigma_2 \mathbf{e}_2$$

so that

$$\mathbf{e}_1 = (y, 0), \qquad \mathbf{e}_2 = (0, y).$$

We have

$$d\sigma_1 = \frac{dx\,dy}{y^2} = \sigma_1\sigma_2, \qquad d\sigma_2 = 0,$$

hence

$$(d\sigma_1, d\sigma_2) = (\sigma_1, \sigma_2)\begin{pmatrix} 0 & \sigma_1 \\ -\sigma_1 & 0 \end{pmatrix},$$

$$\Omega = \begin{pmatrix} 0 & \sigma_1 \\ -\sigma_1 & 0 \end{pmatrix},$$

$$\Theta = d\Omega - \Omega^2 = \begin{pmatrix} 0 & d\sigma_1 \\ -d\sigma_1 & 0 \end{pmatrix} = \begin{pmatrix} 0 & 1 \\ -1 & 0 \end{pmatrix}\sigma_1\sigma_2.$$

The only significant component of the curvature tensor is

$$R_{12\,12} = 1.$$

In our discussion of surfaces we wrote $d\varpi + K\sigma_1\sigma_2 = 0$ for the Gaussian curvature K. Here the right interpretation is $\varpi = \omega_{12} = \sigma_1$, $K = -R_{12\,12} = -1$. For this reason we say that the upper half plane with the Poincaré metric has constant negative curvature.

We shall show that each semicircle

$$P = (a + r\cos t, r\sin t) \qquad 0 < t < \pi$$

orthogonal to the x-axis is a geodesic. In coordinates it is given by

$$x = a + r\cos t, \qquad y = r\sin t,$$

and we have for the tangent

$$\mathbf{d}P/dt = r(-\sin t, \cos t)$$
$$= \frac{r}{y}[(-\sin t)\,\mathbf{e}_1 + (\cos t)\,\mathbf{e}_2]$$
$$= \frac{1}{\sin t}[(-\sin t)\,\mathbf{e}_1 + (\cos t)\,\mathbf{e}_2].$$

Hence for the arc length s,

$$\frac{ds}{dt} = \left[\frac{(-\sin t)^2 + (\cos t)^2}{\sin^2 t}\right]^{1/2} = \frac{1}{\sin t}$$
$$\mathbf{t} = \mathbf{d}P/ds = (-\sin t)\,\mathbf{e}_1 + (\cos t)\,\mathbf{e}_2 .$$

Along the curve we have

$$\omega_{12} = \sigma_1 = -dt,$$
$$\mathbf{de}_1 = -dt\,\mathbf{e}_2 , \qquad \mathbf{de}_2 = dt\,\mathbf{e}_1 ,$$
$$\mathbf{dt} = \mathbf{d}[(-\sin t)\,\mathbf{e}_1 + (\cos t)\,\mathbf{e}_1]$$
$$= -(\cos t)\,dt\,\mathbf{e}_1 - (\sin t)\,dt\,\mathbf{e}_2 - (\sin t)(-dt\,\mathbf{e}_2) + (\cos t)(dt\,\mathbf{e}_1)$$
$$= 0,$$

the unit tangent moves by parallel displacement, the curve is a geodesic.

We shall close this local study with an application of The Frobenius Integration Theorem of Sections 7.3 and 7.4.

Let **M** *be a Riemannian manifold with curvature tensor zero. Then* **M** *is flat: there exists a local coordinate system* $u^1, \cdots, u^n$ *for which the natural frame*

$$\frac{\partial}{\partial u^1}, \cdots, \frac{\partial}{\partial u^n}$$

is an orthonormal frame.

This is proved as follows. We are assuming $\Theta = 0$, i.e., $d\Omega = \Omega^2$. By the first application in Section 7.4, there is a matrix A of functions satisfying

$$(dA)A^{-1} = \Omega,$$

and A is orthogonal. We define $\boldsymbol{\tau} = (\tau_1, \cdots, \tau_n)$ by $\boldsymbol{\tau} = \boldsymbol{\sigma}A$. Then

$$d\boldsymbol{\tau} = d(\boldsymbol{\sigma}A) = (d\boldsymbol{\sigma})\,A - \boldsymbol{\sigma}\,dA = (\boldsymbol{\sigma}\Omega)\,A - \boldsymbol{\sigma}(\Omega A) = 0.$$

Each of the one-forms τ_i is closed, $d\tau_i = 0$, hence exact locally,

$$\tau_i = du^i.$$

This defines our local coordinate system $(u^1, \cdots, u^n)$. On the one hand we have

$$\mathbf{d}P = \sum du^i\left(\frac{\partial}{\partial u^i}\right)$$

and on the other,

$$\mathbf{d}P = \sigma\mathbf{e} = \tau A^{-1}\mathbf{e},$$

hence

$$\begin{pmatrix} \partial/\partial u^1 \\ \vdots \\ \partial/\partial u^n \end{pmatrix} = A^{-1}\mathbf{e}.$$

Since the frame $\mathbf{e}$ is orthonormal and the matrix A is orthogonal, the natural frame $(\partial/\partial u^1, \cdots, \partial/\partial u^n)$ is also orthonormal.

8.4. Riemannian Geometry, Harmonic Integrals

In this section we shall sketch the remarkable results of W. V. D. Hodge on the potential theory of closed Riemannian manifolds. This work pertains to differential forms alone so we can forget all about vector fields. In this spirit we had better make a fresh start and reformulate the pertinent facts about Riemannian manifolds which we shall need. We shall presuppose that the manifolds we discuss are orientable, so this will be built into the structure.

Thus we have a manifold $\mathbf{M}$. It is covered by a system of overlapping neighborhoods $\mathbf{U}_1, \mathbf{U}_2, \cdots$. On each $\mathbf{U}$, there is a basis

$$\sigma_1, \cdots, \sigma_n$$

for the one-forms. If $\sigma_1, \cdots, \sigma_n$ is this basis on $\mathbf{U}$ and $\bar{\sigma}_1, \cdots, \bar{\sigma}_n$ is the one on $\bar{\mathbf{U}}$, then wherever $\mathbf{U}$ and $\bar{\mathbf{U}}$ intersect we must have

$$\bar{\sigma}_i = \sum a_{ij}\sigma_j$$

where $A = \|a_{ij}\|$ is a proper (determinant one) orthogonal matrix.

The volume element $\sigma_1, \cdots, \sigma_n$ is an intrinsic quantity, according to

$$\bar{\sigma}_1 \cdots \bar{\sigma}_n = |a_{ij}|\sigma_1 \cdots \sigma_n = \sigma_1 \cdots \sigma_n.$$

Next, the star operator of Section 2.7 applies. To each p-form ω corresponds on $(n-p)$-form $*\omega$. Locally

$$*(\sigma_1 \cdots \sigma_p) = \sigma_{p+1} \cdots \sigma_n.$$

We recall that

$$**\omega = (-1)^{p(n-p)}\omega.$$

We define a new operator δ by

$$\delta\omega = (-1)^{np+n+1} * d * \omega.$$

The significance of the sign will appear shortly. We note that $\delta\omega$ is a $(p-1)$-form when ω is a p-form. If $\omega = f$ is a zero-form, or function, then $\delta f = 0$. The final operator we define is the harmonic operator (generalized Laplacian) according to

$$\Delta = d \circ \delta + \delta \circ d.$$

We henceforth restrict attention to a closed (compact) manifold $\mathbf{M}$. We denote by τ the volume element, an n-form on $\mathbf{M}$ which nowhere vanishes and which satisfies $\tau = \sigma_1 \cdots \sigma_n$ locally.

We propose to turn the space of p-forms on $\mathbf{M}$ into an (infinite dimensional) inner product space. If ω and η are two p-forms then $\omega \wedge *\eta$ is an n-form and we define

$$(\omega, \eta) = \int_{\mathbf{M}} \omega \wedge *\eta.$$

This is evidently linear in each variable and we have

$$(\omega, \eta) = (\eta, \omega),$$

a consequence of

$$\omega \wedge *\eta = \eta \wedge *\omega.$$

Locally, if

$$\omega = \sum a_H \sigma^H,$$

then

$$\omega \wedge *\omega = (\sum a_H^2)\tau$$

hence

$$(\omega, \omega) \geqq 0$$

and $(\omega, \omega) = 0$ if and only if $\omega = 0$.

We shall now establish the fundamental formula:

If ω is a p-form and η a $(p+1)$-form then

$$(d\omega, \eta) = (\omega, \delta\eta).$$

For we integrate over the closed manifold $\mathbf{M}$ the relation

$$d\omega \wedge *\eta + (-1)^p \omega \wedge d*\eta = d(\omega \wedge *\eta):$$

$$\int_{\mathbf{M}} d\omega \wedge *\eta + (-1)^p \int_{\mathbf{M}} \omega \wedge d*\eta = \int_{\mathbf{M}} d(\omega \wedge *\eta) = \int_{\partial\mathbf{M}} \omega \wedge *\eta = 0,$$

$$(d\omega, \eta) = (-1)^{p-1} \int_{\mathbf{M}} \omega \wedge d*\eta.$$

Since $d*\eta$ is an $(n-p)$-form, we have

$$**(d*\eta) = *(*d*\eta) = (-1)^{p(n-p)} d*\eta$$

so that

$$\begin{aligned}(-1)^{p-1} d*\eta &= (-1)^{p-1}(-1)^{p(n-p)}*(*d*\eta)\\ &= (-1)^{np+1}*(*d*\eta).\end{aligned}$$

But η is a $(p+1)$-form, hence

$$\delta\eta = (-1)^{n(p+1)+n+1} * d * \eta = (-1)^{np+1} * d * \eta,$$

and so

$$(-1)^{p-1} d * \eta = * \delta\eta,$$

$$(d\omega, \eta) = (-1)^{p-1} \int_{\mathbf{M}} \omega \wedge d * \eta$$

$$= \int_{\mathbf{M}} \omega \wedge *(\delta\eta) = (\omega, \delta\eta).$$

A form ω is called *harmonic* provided

$$\Delta\omega = 0.$$

It is clear that if the p-form ω satisfies the two equations $d\omega=0$, $\delta\omega = 0$, then ω is harmonic. The converse is also true. Indeed, if ω is any p-form, then

$$(\Delta\omega, \omega) = (d\delta\omega, \omega) + (\delta d\omega, \omega)$$

$$= (\delta\omega, \delta\omega) + (d\omega, d\omega).$$

Now if ω is harmonic, then $\Delta\omega = 0$,

$$(\delta\omega, \delta\omega) + (d\omega, d\omega) = 0.$$

But each term is nonnegative, hence each vanishes, $(d\omega, d\omega) = 0$, $(\delta\omega, \delta\omega) = 0$ and this implies in turn that $d\omega = 0$, $\delta\omega = 0$.

The operators d, δ, act on the space of p-forms. The relation $(d\omega, \eta) = (\omega, \delta\eta)$ may be interpreted as saying that d and δ are adjoint to each other. We next see that Δ, which maps p-forms into p-forms, is self-adjoint,

$$(\Delta\omega, \eta) = (\omega, \Delta\eta),$$

indeed, either side is $(d\omega, d\eta) + (\delta\omega, \delta\eta)$. Since $(\Delta\omega, \omega) \geqq 0$ with equality only when $\Delta\omega = 0$, we are entitled to call Δ a positive definite (or elliptic) self-adjoint differential operator.

We may now state Hodge's main result, a deep theorem in harmonic analysis:

If ω is any p-form then there is a $(p-1)$-form α, a $(p+1)$-form β and a harmonic p-form γ such that

$$\omega = d\alpha + \delta\beta + \gamma.$$

The forms $d\alpha$, $\delta\beta$, γ are unique.

The proof that α, β, and γ exist is difficult. We shall only settle the uniqueness part. Suppose we have

$$d\alpha + \delta\beta + \gamma = 0.$$

We then have $d(d\alpha) = 0$ and also $d\gamma = 0$ since γ is harmonic. Hence

$$\begin{aligned} d\delta\beta &= 0, \\ (d\delta\beta, \beta) &= 0, \\ (\delta\beta, \delta\beta) &= 0, \\ \delta\beta &= 0, \\ d\alpha + \gamma &= 0. \end{aligned}$$

Similarly $d\alpha = 0$, $\gamma = 0$.

By an almost identical argument one shows that in case ω is a closed p-form, $d\omega = 0$, then the term $\delta\beta$ in the Hodge decomposition of ω is absent.

$$\omega = d\alpha + \gamma.$$

It follows from this that if $\mathbf{z}$ is any p-cycle, then

$$\int_{\mathbf{z}} \omega = \int_{\mathbf{z}} \gamma,$$

that is, γ has the same periods as does ω. (See De Rham's theorems, Section 5.9.) The result of this is that *if ω is any closed form, then there exists a unique harmonic form γ with the same periods as those of ω.*

We can also answer the following question. Given a p-form λ, when is there a p-form η such that the equation

$$\Delta\eta = \lambda$$

is satisfied? The answer is: if and only if

$$(\gamma, \lambda) = 0$$

for every harmonic form γ.

For suppose $\lambda = \Delta\eta$ and γ is harmonic. Then

$$(\gamma, \lambda) = (\gamma, \Delta\eta) = (\Delta\gamma, \eta) = (0, \eta) = 0.$$

On the other hand, suppose λ is a form satisfying $(\gamma, \lambda) = 0$ for each harmonic form γ. From the decomposition

$$\lambda = d\alpha + \delta\beta + \gamma$$

we have, using the particular γ which is part of λ,

$$\begin{aligned} 0 = (\gamma, \lambda) &= (\gamma, d\alpha) + (\gamma, \delta\beta) + (\gamma, \gamma) \\ &= (\delta\gamma, \alpha) + (d\gamma, \beta) + (\gamma, \gamma) \\ &= (\gamma, \gamma), \end{aligned}$$

hence $\gamma = 0$,

$$\lambda = d\alpha + \delta\beta.$$

We shall set $\eta = \mu + \nu$ and try to solve $\Delta\mu = d\alpha$, $\Delta\nu = \delta\beta$, separately. We take the first

$$\Delta\mu = d\alpha.$$

Decomposing α,

$$\alpha = d\alpha_1 + \delta\beta_1 + \gamma_1,$$
$$d\alpha = d\,\delta\beta_1.$$

Next,

$$\beta_1 = d\alpha_2 + \delta\beta_2 + \gamma_2,$$
$$d\,\delta\beta_1 = d\delta\, d\alpha_2 = (d\delta + \delta d)(d\alpha_2) = \Delta(d\alpha_2),$$
$$d\alpha = \Delta\mu \qquad \text{with} \qquad \mu = d\alpha_2.$$

We find ν similarly.

Example 1. $\mathbf{E}^n$. We shall compute the operator Δ in $\mathbf{E}^n$. Contrary to our previous notation in dealing with the standard Laplacian, we shall denote the Laplacian by Lap,

$$\operatorname{Lap} u = \sum u_{ii} = \sum \frac{\partial^2 u}{\partial x^i\, \partial x^i}.$$

The result is this: if

$$\omega = \sum a_H\, dx^H,$$

then

$$\Delta\omega = -\sum (\operatorname{Lap} a_H)\, dx^H.$$

It will suffice to establish this for the monomial

$$\omega = A\, dx^1 \cdots dx^p.$$

We shall abbreviate the calculation by these conventions:

(1) subscripts on A denote partial derivatives;
(2) $\alpha = 1, 2, \cdots, p$, $j = p + 1, \cdots, n$;
(3) each repeated index is summed over its range.

We also remark that in taking the star of a monomial, the choice of sign is always governed by the rule $\eta \wedge *\eta = B\, dx^1 \cdots dx^n$ where $B > 0$, for example,

$$*(dx^\alpha\, dx^{p+1} \cdots \widehat{dx^j} \cdots dx^n) = (-1)^{np+p+\alpha+j}\, dx^1 \cdots \widehat{dx^\alpha} \cdots dx^p\, dx^j.$$

Another point: since ω is a p-form and $d\omega$ is a $(p+1)$-form,

$$\delta\omega = (-1)^{np+n+1} * d * \omega,$$
$$\delta d\omega = (-1)^{np+1} * d * d\omega.$$

We have

$$*\omega = A\, dx^{p+1} \cdots dx^n,$$

$$d*\omega = A_\alpha\, dx^\alpha\, dx^{p+1} \cdots dx^n,$$

$$*d*\omega = (-1)^{np+n+\alpha+1} A_\alpha\, dx^1 \cdots \widehat{dx^\alpha} \cdots dx^p,$$

$$\delta\omega = (-1)^\alpha A_\alpha\, dx^1 \cdots \widehat{dx^\alpha} \cdots dx^p,$$

$$d\,\delta\omega = -A_{\alpha\alpha}\, dx^1 \cdots dx^p + (-1)^{\alpha+p+1} A_{\alpha j}\, dx^1 \cdots \widehat{dx^\alpha} \cdots dx^p\, dx^j.$$

Next,

$$d\omega = A_j\, dx^j\, dx^1 \cdots dx^p,$$

$$*d\omega = (-1)^{j+1} A_j\, dx^{p+1} \cdots \widehat{dx^j} \cdots dx^n,$$

$$d*d\omega = (-1)^p A_{jj}\, dx^{p+1} \cdots dx^n + (-1)^{j+1} A_{j\alpha}\, dx^\alpha\, dx^{p+1} \cdots \widehat{dx^j} \cdots dx^n,$$

$$*d*d\omega = (-1)^{np} A_{jj}\, dx^1 \cdots dx^p + (-1)^{np+p+\alpha+1} A_{j\alpha}\, dx^1 \cdots \widehat{dx^\alpha} \cdots dx^p\, dx^j,$$

$$\delta\, d\omega = -A_{jj}\, dx^1 \cdots dx^p + (-1)^{p+\alpha} A_{j\alpha}\, dx^1 \cdots \widehat{dx^\alpha} \cdots dx^p\, dx^j.$$

Combining these expressions,

$$\begin{aligned} \Delta\omega = d\delta\omega + \delta d\omega &= -[A_{\alpha\alpha} + A_{jj}]\, dx^1 \cdots dx^p \\ &= -(\operatorname{Lap} A)\, dx^1 \cdots dx^p. \end{aligned}$$

The minus sign seems strange in view of the relation

$$(\Delta\omega, \omega) \geqq 0$$

on closed manifolds. An example may clear this point. Let us take the zero-form $\sin x$ on the flat torus $0 \leqq x, y, z \leqq 2\pi$, where numbers are identified if they differ by a multiple of 2π. Then

$$\begin{aligned} d(\sin x) &= (\cos x)\, dx \\ *d(\sin x) &= (\cos x)\, dy\, dz \\ d*d(\sin x) &= -(\sin x)\, dx\, dy\, dz, \\ *d*d(\sin x) &= -(\sin x), \\ \Delta(\sin x) &= \delta\, d(\sin x) = +\sin x, \end{aligned}$$

$$(\Delta(\sin x), \sin x) = (2\pi)^2 \int_0^{2\pi} \sin^2 x\, dx = (2\pi)^2\pi > 0.$$

We have used the fact that $d(\sin x)$ is a one-form, $n = 3, p = 1$,

$$\delta[d(\sin x)] = (-1)^{np+n+1}(* d*)d(\sin x) = -*d*d(\sin x).$$

Example 2. $\mathbf{S}^2$. If f is a function on $\mathbf{E}^3$, we have the spherical coordinate form of the Laplacian (Section 4.4, p. 40),

$$\operatorname{Lap} f = \frac{1}{r^2 \sin\phi}\left[\frac{\partial}{\partial r}\left(r^2 \sin\phi \frac{\partial f}{\partial r}\right) + \frac{\partial}{\partial \phi}\left(\sin\phi \frac{\partial f}{\partial \phi}\right) + \frac{\partial}{\partial \theta}\left(\frac{1}{\sin\phi}\frac{\partial f}{\partial \theta}\right)\right].$$

Suppose that $\operatorname{Lap} f = 0$ and that f is of the form

$$f(r, \phi, \theta) = r^n g(\phi, \theta).$$

Then g is called a *spherical harmonic* and must satisfy

$$n(n+1)(\sin\phi)g + \frac{\partial}{\partial\phi}[(\sin\phi)g_\phi] + \frac{\partial}{\partial\theta}\left(\frac{g_\theta}{\sin\phi}\right) = 0.$$

The function g may be considered as a function, or zero-form, on $\mathbf{S}^2$, the unit sphere. There we have

$$\sigma_1 = d\phi, \qquad \sigma_2 = \sin\phi\, d\theta,$$

$$dg = g_\phi \sigma_1 + \left(\frac{g_\theta}{\sin\phi}\right)\sigma_2,$$

$$*dg = \left(\frac{-g_\theta}{\sin\phi}\right)\sigma_1 + g_\phi \sigma_2$$

$$= -\left(\frac{g_\theta}{\sin\phi}\right)d\phi + (\sin\phi)g_\phi\, d\theta,$$

$$d*dg = \frac{\partial}{\partial\theta}\left(\frac{g_\theta}{\sin\phi}\right)d\phi\, d\theta + \frac{\partial}{\partial\phi}[(\sin\phi)g_\phi]\, d\phi\, d\theta$$

$$= \frac{1}{\sin\phi}\left[\frac{\partial}{\partial\theta}\left(\frac{g_\theta}{\sin\phi}\right) + \frac{\partial}{\partial\phi}((\sin\phi)g_\phi)\right]\sigma_1\sigma_2,$$

$$*d*dg = \frac{1}{\sin\phi}\left[\frac{\partial}{\partial\theta}\left(\frac{g_\theta}{\sin\phi}\right) + \frac{\partial}{\partial\phi}(g_\phi \sin\phi)\right].$$

But

$$\Delta g = (d\delta + \delta d)g = \delta\, dg = -*d*dg,$$

so the condition for g to be a spherical harmonic is

$$\Delta g - n(n+1)g = 0.$$

In other words, the spherical harmonics are eigenfunctions for the generalized Laplacian on $\mathbf{S}^2$. Many of the usual facts about spherical harmonics follow from our calculations. We take one example, the orthogonality relation: *If g and h are spherical harmonics of distinct degrees m and n, respectively, then* $(g, h) = 0$.

For $\Delta g = m(m + 1)g$, $\Delta h = n(n + 1)h$, hence

$$(g, h) = \frac{1}{m(m+1)}(\Delta g, h) = \frac{1}{m(m+1)}(g, \Delta h)$$

$$= \frac{n(n+1)}{m(m+1)}(g, h),$$

$$(g, h) = 0.$$

8.5. Affine Connection

We shall approach the problem of affine connection this way. We seek the weakest structure with which we can endow a manifold so that parallel displacement of vectors along curves is possible. Considerably less than a Riemannian structure is required.

Let $\mathbf{M}$ be a manifold. An *affine frame* (or simply *frame*) on a neighborhood $\mathbf{U}$ of $\mathbf{M}$ consists of n vector fields $\mathbf{e}_1, \cdots, \mathbf{e}_n$ on $\mathbf{U}$ which are linearly independent at each point of $\mathbf{U}$. Thus at each point P of $\mathbf{U}$ the vectors $(\mathbf{e}_1)_P, \cdots, (\mathbf{e}_n)_P$ furnish a basis of the tangent space $\mathbf{T}$ at P. There is a dual basis $\sigma^1, \cdots, \sigma^n$ of one-forms on $\mathbf{U}$ so we may write

$$\mathbf{d}P = \sum \sigma^i \mathbf{e}_i$$

as we did for the Riemannian case in Section 8.3.

We now want to associate with each vector field $\mathbf{v}$ on $\mathbf{M}$ a vector field $\mathbf{dv}$ with one-form coefficients. We must be able to do this for the vector fields $\mathbf{e}_i$ of the basis, so we require

$$\mathbf{de}_i = \sum \omega_i{}^j \mathbf{e}_j,$$

where the $\omega_i{}^j$ are one-forms on the neighborhood $\mathbf{U}$. There are certain consistency conditions which guarantee that the computation of $\mathbf{dv}$ will be independent of any frame.

Locally we may describe an affine connection as follows. We are given $\mathbf{U}$, the affine frame $\mathbf{e}_1, \cdots, \mathbf{e}_n$, and the dual basis $\sigma^1, \cdots, \sigma^n$ of one-forms. An affine connection consists of n^2 one-forms $\omega_i{}^j$ subject to no constraints whatever.

We shall develop some of the local geometry of an affine connection before attacking the crucial problem of finding proper consistency conditions which make the definition of an affine connection over a whole manifold possible.

By introducing matrix notation:

$$\mathbf{e} = \begin{pmatrix} \mathbf{e}_1 \\ \vdots \\ \mathbf{e}_n \end{pmatrix}, \qquad \sigma = (\sigma^1, \cdots, \sigma^n), \qquad \Omega = \begin{pmatrix} \omega_{11} & \cdots & \omega_{1n} \\ \vdots & & \\ \omega_{n1} & \cdots & \omega_{nn} \end{pmatrix},$$

we may write our basic structure equations:

$$\mathbf{d}P = \sigma\mathbf{e}, \qquad \mathbf{de} = \Omega\mathbf{e}.$$

We shall quickly point out the relation to the customary tensor formulation of an affine connection. First we expand each $\omega_i{}^j$ in the basis σ^k:

$$\omega_i{}^j = \sum \Gamma_i{}^j{}_k \sigma^k,$$

defining the *connection coefficients* $\Gamma_i{}^j{}_k$. In the usual tensor formulation, the frame $\mathbf{e}_1, \cdots, \mathbf{e}_n$ stems from local coordinates:

$$\mathbf{e}_i = \frac{\partial}{\partial u^i},$$

and correspondingly, $\sigma^i = du^i$. The $\Gamma_i{}^j{}_k = \Gamma_i{}^j{}_k(u^1, \cdots, u^n)$ are then n^3 arbitrary functions assigned on $\mathbf{U}$.

We derive relations by differentiating the structure equations several times. First

$$\mathbf{d}^2 P = (d\sigma)\,\mathbf{e} - \sigma\,\mathbf{de} = (d\sigma - \sigma\Omega)\,\mathbf{e} = \boldsymbol{\tau}\mathbf{e}.$$

Here

$$\boldsymbol{\tau} = (\tau^1, \cdots, \tau^n) = d\sigma - \sigma\Omega.$$

The two-forms τ^i are the *torsion forms.* We may write

$$\tau^i = \tfrac{1}{2} \sum T^i{}_{jk} \sigma^j \wedge \sigma^k, \qquad T^i{}_{jk} + T^i{}_{kj} = 0,$$

defining the *torsion coefficients* $T^i{}_{jk}$. Next,

$$\mathbf{d}^2\mathbf{e} = (d\Omega)\,\mathbf{e} - \Omega\mathbf{de} = (d\Omega - \Omega^2)\,\mathbf{e} = \Theta\mathbf{e}.$$

Here

$$\Theta = d\Omega - \Omega^2 = \|\theta_i{}^j\|$$

is the matrix of *curvature forms* $\theta_i{}^j$. The *curvature tensor* $R_i{}^j{}_{kl}$ is obtained from

$$\theta_i{}^j = \tfrac{1}{2} \sum R_i{}^j{}_{kl} \sigma^k \wedge \sigma^l, \qquad R_i{}^j{}_{kl} + R_i{}^j{}_{lk} = 0.$$

Integrability conditions are obtained by applying the exterior derivative to the equations $\boldsymbol{\tau} = d\sigma - \sigma\Omega$ and $\Theta = d\Omega - \Omega^2$:

$$\begin{aligned} d\tau &= (-d\sigma)\,\Omega + \sigma d\Omega \\ &= -(\tau + \sigma\Omega)\,\Omega + \sigma(\Theta + \Omega^2), \\ d\tau &= \sigma\Theta - \tau\Omega, \end{aligned}$$

and

$$\begin{aligned} d\Theta &= (-d\Omega)\,\Omega + \Omega\, d\Omega \\ &= -(\Theta + \Omega^2)\,\Omega + \Omega(\Theta + \Omega^2), \\ d\Theta &= \Omega\Theta - \Theta\Omega. \end{aligned}$$

If **v** is a vector field, then

$$\mathbf{v} = \sum f^i \mathbf{e}_i = F\mathbf{e}, \qquad F = (f^1, \cdots, f^n),$$

where the f^i are scalars; we have

$$\mathbf{dv} = (dF)\,\mathbf{e} + F(\mathbf{de}) = (dF + F\Omega)\mathbf{e}.$$

Suppose the vector field **v** is defined over a submanifold. It is said to *move by parallel displacement* if $\mathbf{dv} = 0$, i.e.,

$$dF + F\Omega = 0.$$

If $P = P(t)$ is a smooth curve on **U** defined over an interval $t_0 \leqq t \leqq t_1$ of the t axis, if $\mathbf{v}_0$ is a tangent vector at $P_0 = P(t_0)$, then there exists a unique assignment of a tangent vector $\mathbf{v}(t)$ at $P(t)$ for each value of t such that $\mathbf{v}(t_0) = \mathbf{v}_0$ and $\mathbf{v}(t)$ moves by parallel displacement.

For we write $\mathbf{v} = \sum f^i \mathbf{e}_i$. Along the curve the conditions for parallel displacement become

$$\frac{df^j}{dt} + \sum f^i \cdot \left(\frac{\omega_i{}^j}{dt}\right) = 0,$$

a first order linear system which taken with the initial data determines the f^i uniquely.

Now we tackle the global situation. We have a manifold **M** in front of us and we must consider each conceivable moving affine frame $\mathbf{e}_1, \cdots, \mathbf{e}_n$ together with its neighborhood of definition **U**. With each one of these **e** we have an affine connection, i.e., an $n \times n$ matrix Ω of one-forms on **U**. We want these to "fit together" whenever two such neighborhoods overlap.

Thus let $(\mathbf{U}, \mathbf{e}, \sigma, \Omega)$ be one such system and $(\bar{\mathbf{U}}, \bar{\mathbf{e}}, \bar{\sigma}, \bar{\Omega})$ another, where we assume the neighborhoods **U** and $\bar{\mathbf{U}}$ overlap. The basic thing we require is that for any vector field **v** defined on the intersection of **U** and $\bar{\mathbf{U}}$, the computation for **dv** in either connection must yield the same result.

On the overlap, where we shall always operate in what follows,

$$\bar{\mathbf{e}} = A\mathbf{e}$$

where $A = \|a_i{}^j\|$ is a nonsingular matrix of functions. Since

$$\mathbf{d}P = \bar{\sigma}\bar{\mathbf{e}} = \sigma\mathbf{e},$$

we have

$$\bar{\sigma}A\mathbf{e} = \sigma\mathbf{e},$$

$$\bar{\sigma} = \sigma A^{-1}.$$

Now

$$\begin{aligned}\mathbf{d}\bar{\mathbf{e}} &= \mathbf{d}(A\mathbf{e}) = (dA)\,\mathbf{e} + A(\mathbf{de})\\ &= (dA + A\Omega)\,\mathbf{e}\\ &= (dA + A\Omega)\,A^{-1}\bar{\mathbf{e}}.\end{aligned}$$

But also $\mathbf{d\bar{e}} = \bar{\Omega}\mathbf{\bar{e}}$, so we have the transformation law for Ω:

$$\bar{\Omega} = A\Omega A^{-1} + (dA)A^{-1}.$$

This is quite different from the transformation law for σ which is forced by the very definition of manifold and frame. It tells us how the various matrices Ω we are associating with the various moving frames $\mathbf{e}$ must be related if we are to define an affine connection on $\mathbf{M}$ as a whole.

From these formulas one can derive the transformation laws for τ and Θ:

$$\bar{\tau} = \tau A^{-1},$$

$$\bar{\Theta} = A\Theta A^{-1}.$$

If $\mathbf{v} = \bar{F}\mathbf{\bar{e}} = F\mathbf{e}$ is a vector field, then $\bar{F} = FA^{-1}$,

$$(d\bar{F} + \bar{F}\bar{\Omega})\mathbf{\bar{e}}$$

$$= (dF\,A^{-1} - FA^{-1}dA\,A^{-1})A\mathbf{e} + (FA^{-1})(A\Omega A^{-1} + dA\,A^{-1})A\mathbf{e}$$

$$= (dF + F\Omega)\,\mathbf{e},$$

so that $\mathbf{dv}$ is the same, either way it is computed. (We have used the rule $d(A^{-1}) = -A^{-1}(dA)A^{-1}$, as we shall several times.)

We shall now have a second look, only in our present context, at the considerations of Section 4.3. This will provide us with another way of looking at affine connection. What we shall do fits into a general pattern: *quantities subject to a transformation law become absolute invariants when considered on a suitably extended space.*

We begin with a manifold $\mathbf{M}$ of n dimensions. We form a new manifold $\mathbf{F}$ of dimension $(n + n^2)$. This *frame manifold* consists of all frames at all points of $\mathbf{M}$. Precisely, at each point P of $\mathbf{M}$ consider all possible bases $\mathbf{e}_1, \cdots, \mathbf{e}_n$ of the tangent space $\mathbf{T}_P$ at P, and do this for all P.

We obtain coordinates on $\mathbf{F}$ this way. First let $\mathbf{U}$ be a local coordinate neighborhood on $\mathbf{M}$ with coordinates $u^1, \cdots, u^n$. Pick a moving frame $\mathbf{e}_1, \cdots, \mathbf{e}_n$ on $\mathbf{F}$. Thus at each point P of $\mathbf{U}$, $(\mathbf{e}_1)_P, \cdots, (\mathbf{e}_n)_P$ is one basis of the tangent space $\mathbf{T}_P$ at P. The most general basis of $\mathbf{T}_P$ stems from this given one by applying an arbitrary nonsingular transformation. It is $\mathbf{f}_1, \cdots, \mathbf{f}_n$ where

$$\mathbf{f}_i = \sum b_i^{\,j}\cdot(\mathbf{e}_j)_P$$

with $\|b_i^{\,j}\|$ an arbitrary $n \times n$ nonsingular matrix. It is clear from this that the $(n + n^2)$ independent variables

$$u^1, \cdots, u^n, b_i^{\,j}$$

serve as a coordinate system for the neighborhood in $\mathbf{F}$ consisting of all frames at all points P of $\mathbf{U}$.

With the moving frame $\mathbf{e}_1, \cdots, \mathbf{e}_n$ on $\mathbf{U}$ goes the dual basis $\sigma^1, \cdots, \sigma^n$ of one-forms on $\mathbf{U}$. The forms $\tilde{\sigma}^1, \cdots, \tilde{\sigma}^n$ defined by

$$\tilde{\boldsymbol{\sigma}} = \boldsymbol{\sigma} B^{-1},$$

where

$$\boldsymbol{\sigma} = (\sigma^1, \cdots, \sigma^n), \qquad \tilde{\boldsymbol{\sigma}} = (\tilde{\sigma}^1, \cdots, \tilde{\sigma}^n), \qquad B = \|b_i{}^j\|,$$

are one-forms on the part of $\mathbf{F}$ lying over $\mathbf{U}$. More precisely, the values of the forms $\tilde{\sigma}^1, \cdots$ at the point $\mathbf{f}$ of $\mathbf{F}$ given by

$$\mathbf{f}_i = \sum b_i{}^j (\mathbf{e}_j)_P$$

are

$$\tilde{\boldsymbol{\sigma}}|_{\mathbf{f}} = (\boldsymbol{\sigma}\,|_P) B^{-1}.$$

Now suppose that $\bar{\mathbf{U}}$ is a second coordinate neighborhood, $\bar{\mathbf{e}}$ a moving frame on $\bar{\mathbf{U}}$, etc.. Suppose that P lies both in $\mathbf{U}$ and in $\bar{\mathbf{U}}$. We have

$$\bar{\mathbf{e}} = A\mathbf{e}$$

where A is a matrix of functions,

$$\bar{\boldsymbol{\sigma}} = \boldsymbol{\sigma} A^{-1},$$

all worked out above. If the point $\mathbf{f}$ of $\mathbf{F}$ has coordinates B with respect to $\mathbf{e}$ and $\bar{B}$ with respect to $\bar{\mathbf{e}}$, then

$$B\mathbf{e} = \bar{B}\bar{\mathbf{e}} = \bar{B}A\mathbf{e},$$

$$B = \bar{B}A.$$

Thus

$$\tilde{\boldsymbol{\sigma}} = \boldsymbol{\sigma} B^{-1} = (\bar{\boldsymbol{\sigma}} A)(\bar{B}A)^{-1} = \bar{\boldsymbol{\sigma}}\,\bar{B}^{-1}.$$

This implies that *the one-forms $\tilde{\sigma}^1, \cdots, \tilde{\sigma}^n$ are defined on all of* $\mathbf{F}$ *and are completely independent of the particular local coordinate neighborhoods and moving frames used in their definitions.*

This is of first importance. We began with an n-manifold $\mathbf{M}$. We constructed over it a new manifold $\mathbf{F}$. On this new manifold we automatically have, free of charge, the n (linearly independent) one-forms $\tilde{\sigma}^1, \cdots, \tilde{\sigma}^n$.

Now suppose an affine connection is given on $\mathbf{M}$. Thus to the neighborhood $\mathbf{U}$ with a definite affine frame $\mathbf{e}$ is given a matrix $\Omega = \|\omega_i{}^j\|$ of one-forms. The matrix $\bar{\Omega}$ corresponding to the frame $\bar{\mathbf{e}}$ on $\bar{\mathbf{U}}$ is related to Ω (on the overlap of $\mathbf{U}$ and $\bar{\mathbf{U}}$) by a certain transformation law. This transformation law means nothing more nor less than that the n^2 one-forms $\tilde{\omega}_i{}^j$ defined by

$$\tilde{\Omega} = \|\tilde{\omega}_i{}^j\| = B\Omega B^{-1} + (dB)B^{-1},$$

apparently defined only on the part of $\mathbf{F}$ lying over $\mathbf{U}$, are defined on all of $\mathbf{F}$, are completely independent of $\mathbf{e}$ and its $\mathbf{U}$.

This statement may be verified by a calculation. According to the notation above, one must prove the formula

$$B\Omega B^{-1} + (dB)\,B^{-1} = \bar{B}\bar{\Omega}\bar{B}^{-1} + (d\bar{B})\,\bar{B}^{-1}.$$

We leave the details to the reader, but point out that the result can be motivated by a symbolic calculation:

$$\begin{aligned}\mathbf{f} &= B\mathbf{e},\\ d\mathbf{f} &= (dB)\,\mathbf{e} + B(d\mathbf{e}) = (dB + B\Omega)\,\mathbf{e}\\ &= (dB + B\Omega)\,B^{-1}\mathbf{f},\end{aligned}$$

similarly

$$d\mathbf{f} = (d\bar{B} + \bar{B}\bar{\Omega})\,\bar{B}^{-1}\mathbf{f},$$

hence

$$(dB + B\Omega)\,B^{-1} = (d\bar{B} + \bar{B}\bar{\Omega})\,\bar{B}^{-1}.$$

From the one-forms $\tilde{\sigma}^1, \cdots, \tilde{\omega}_i{}^j$ on $\mathbf{F}$, one constructs two-forms $\tilde{\tau}^i, \tilde{\theta}_i{}^j$ according to

$$\begin{aligned}\tilde{\boldsymbol{\tau}} &= (\tilde{\tau}^1, \cdots, \tilde{\tau}^n) = d\tilde{\boldsymbol{\sigma}} - \tilde{\boldsymbol{\sigma}}\tilde{\Omega},\\ \tilde{\Theta} &= \|\tilde{\theta}_i{}^j\| = d\tilde{\Omega} - \tilde{\Omega}^2.\end{aligned}$$

These are the *general torsion* and *curvature forms* respectively.

8.6. Problems

1. Let Σ be a closed convex surface with constant mean curvature H. Prove that Σ is a sphere.

2. A surface Σ is given in the Monge form $z = f(x, y)$, defined for all x, y. We suppose this surface is convex from below. Show that this means that

$$\begin{pmatrix} r & s \\ s & t \end{pmatrix}$$

is positive semidefinite.

3. (Continuation.) Show that the mapping

$$(x, y) \longrightarrow (x + p, y + q)$$

increases distance and hence is one-one.

4. A point of a hypersurface is an *umbilic* if the transformation A at that point has all of its characteristic roots (principal curvatures) equal. Let $\mathbf{M}$ be a hypersurface all of whose points are umbilics. Prove that $\mathbf{M}$ is a portion of a hyperplane or sphere.

5. Suppose on a Riemannian manifold $\mathbf{M}$ there is a scalar K such that

$$\theta_{ij} = -K\sigma_i \wedge \sigma_j.$$

Prove that K is a constant.

6. Let **M** be a manifold with an affine connection given. Show that the two-form α defined locally by

$$\alpha = \sum \theta_i{}^i = \text{trace}\, \Theta$$

is actually defined on all of **M**, independent of local frames. Show also that $d\alpha = 0$. It is even true that there exists a one-form λ on **M** such that $\alpha = d\lambda$, but this is difficult to establish.

7. (Continuation.) Prove

$$d\Theta^r = \Omega\Theta^r - \Theta^r\Omega$$

and that

$$d[\text{trace}\,(\Theta^r)] = 0.$$

8. Let **M** be a manifold with affine connection. Given an affine frame $\mathbf{e}_1, \cdots, \mathbf{e}_n$ with corresponding connection coefficients Γ and torsion coefficients T, define a new connection by specifying the new connection coefficients Γ^*:

$$\Gamma^*{}_i{}^j{}_k = \Gamma_i{}^j{}_k + \tfrac{1}{2}T^j{}_{ik}\,.$$

Show that this indeed defines a connection on <u>all</u> of **M** and that this connection is *symmetric* (no torsion). Investigate the meaning of "symmetric" for a local coordinate frame.

9. Consider the flat torus $\mathbf{T}^n$ which consists of all points $(x_1, \cdots, x_n)$ where each x_i is taken modulo one. That is, $(x_1, \cdots, x_n) = (y_1, \cdots, y_n)$ if $x_i - y_i = \text{integer}$ $(i = 1, \cdots, n)$. The flat metric is given by the orthonormal basis

$$\sigma_1 = dx_1, \cdots, \qquad \sigma_n = dx_n$$

of one-forms. (In older notation, $[ds]^2 = [dx_1]^2 + \cdots + [dx_n]^2$.) Find all harmonic differentials of all degrees.

IX

Applications to Group Theory

9.1. Lie Groups

A *Lie group* consists of a smooth manifold $\mathbf{G}$ which has a group structure

$$(x, y) \longrightarrow xy.$$

We suppose that this group operation, which may be considered as a mapping

$$\mathbf{G} \times \mathbf{G} \longrightarrow \mathbf{G},$$

is smooth and also that the map $x \longrightarrow x^{-1}$ on $\mathbf{G} \longrightarrow \mathbf{G}$ is smooth.

With each element x in $\mathbf{G}$, there is associated a transformation L_x of $\mathbf{G}$, called *left translation*:

$$L_x(y) = xy.$$

A differential p-form ω is called *left invariant* provided

$$L_x{}^* \omega = \omega$$

for all x in $\mathbf{G}$.

Let e denote the unit element of $\mathbf{G}$. The left translation $L_x{}^{-1} = L_{x^{-1}}$ sends x to e. If ω is left invariant, $\omega = L^*_{x^{-1}}\,\omega$ is completely determined at x by its value ω_0 at e. If ω_0 is any given p-form at e, then a left invariant form ω is defined by

$$\omega_x = L^*_{x^{-1}}\,\omega_0 .$$

These remarks serve to determine the existence of left invariant forms. Let us begin with one-forms. We let n be the dimension of $\mathbf{G}$. Since the space of one-forms at e is an n-dimensional linear space, there are exactly n linearly independent left invariant one-forms on $\mathbf{G}$. Let

$$\sigma^1, \cdots, \sigma^n$$

be such a system. Any other left invariant one-form is a linear combination of these with constant coefficients.

More generally, if ω is any left invariant p-form on $\mathbf{G}$,

$$\omega = \sum c_H \sigma^H,$$

where the c_H are constants and $\sigma^H = \sigma^{h_1} \cdots \sigma^{h_p}$. Any p-form ω can be expanded in this way and the coefficients c_H will, in general, be scalars on **G**. Supposing ω left invariant forces each of these scalars to be left invariant. This means that each c_H takes the same value at each point of **G**, hence is constant.

Next, if ω is left invariant, so is $d\omega$, since

$$L_x{}^*(d\omega) = d(L_x{}^*\,\omega) = d\omega.$$

It follows that there are *constants of structure* $c^i{}_{jk}$ such that

$$d\sigma^i = \tfrac{1}{2} \sum c^i{}_{jk}\sigma^j\sigma^k, \qquad c^i{}_{jk} + c^i{}_{kj} = 0.$$

Substituting this into the relations $d(d\sigma^i) = 0$ eventually yields

$$\sum (c^i{}_{jk}c^j{}_{rs} + c^i{}_{jr}c^j{}_{sk} + c^i{}_{js}c^j{}_{kr}) = 0.$$

Particularly important is the n-form

$$\sigma^1 \cdots \sigma^n$$

which defines a left invariant volume element on **G**. It is clear from this that **G** is orientable.

The *right translation* R_z associated to a group element z is the mapping

$$y \longrightarrow R_z y = yz$$

on **G** to **G**. From the associative law

$$x(yz) = (xy)z$$

we deduce that

$$L_x \circ R_z = R_z \circ L_x,$$

hence

$$R_z{}^* \circ L_x{}^* = L_x{}^* \circ R_z{}^*.$$

Suppose ω is a left invariant p-form. Then for each x and z,

$$L_x{}^*(R_z{}^*\,\omega) = R_z{}^*(L_x{}^*\,\omega) = R_z{}^*\,\omega,$$

hence $R_z{}^*\,\omega$ is also a left invariant p-form.

9.2. Examples of Lie Groups

Example 1. $n = 1$. We shall determine the local structure of all one-dimensional groups. Let t be a parameter on **G**, chosen so that $t = 0$ is the identity e. Let σ be a nontrivial left invariant one-form; locally,

$$\sigma = f(t)\,dt, \qquad \text{never zero.}$$

We integrate σ to get a new parameter for **G**,

$$\int_0^t f(t)\,dt.$$

Thus we may assume we have started with a parameterization of a neighborhood of e by a single variable t such that

$$\sigma = dt$$

is a left invariant form.

We next express the group product analytically. The product of the point with coordinate s with that of coordinate t will have coordinate u given by

$$u = p(s, t)$$

with

$$p(s, 0) = s, \qquad p(0, t) = t$$

according to $xe = x$, $ey = y$. In coordinates,

$$L_s\colon \quad t \longrightarrow u = p(s, t).$$

The left invariance of σ, $L_s{}^* \sigma = \sigma$, means

$$dt = \frac{\partial p}{\partial t}\, dt,$$

hence

$$\frac{\partial p}{\partial t} = 1, \qquad p(s, t) = t + \phi(s).$$

Setting $t = 0$:

$$s = p(s, 0) = \phi(s),$$

and so

$$p(s, t) = s + t.$$

It follows that the group operation corresponds to nothing more than ordinary addition of coordinates. As a corollary, **G** is abelian (commutative).

Example 2. **G** is a group for which the constants of structure $c^i{}_{jk}$ all vanish. Thus

$$d\sigma^1 = \cdots = d\sigma^n = 0.$$

In a small neighborhood of e,

$$\sigma^i = du^i,$$

taken so that $(0, 0, \cdots, 0) \longleftrightarrow e$. The product of points with coordinates $\mathbf{u}$, $\mathbf{v}$, respectively, is a point with coordinates $\mathbf{w}$ given by

$$w^i = p^i(\mathbf{u}, \mathbf{v})$$

with

$$p^i(\mathbf{u}, 0) = u^i, \qquad p^i(0, \mathbf{v}) = v^i.$$

The invariance of σ^i under the left translation

$$\mathbf{v} \longrightarrow \mathbf{w}, \qquad w^i = p^i(\mathbf{u}, \mathbf{v})$$

is expressed analytically by

$$dv^i = \sum \frac{\partial p^i}{\partial v^j}(\mathbf{u}, \mathbf{v})\, dv^j$$

which implies

$$\left\| \frac{\partial p^i}{\partial v^j} \right\| = I,$$

$$\frac{\partial p^i}{\partial v^j}(\mathbf{u}, \mathbf{v}) = \delta^i_j,$$

$$p^i(\mathbf{u}, \mathbf{v}) = v^i + \phi^i(\mathbf{u}).$$

Setting $\mathbf{v} = 0$ yields

$$u^i = \phi^i(\mathbf{u}),$$

hence

$$p^i(\mathbf{u}, \mathbf{v}) = u^i + v^i.$$

If $P(\mathbf{u})$ denotes the point with coordinates $\mathbf{u}$, this says

$$P(\mathbf{u}) \cdot P(\mathbf{v}) = P(\mathbf{u} + \mathbf{v})$$

so that locally the group looks like a neighborhood of 0 in $\mathbf{E}^n$.

Corollary. $\mathbf{G}$ is abelian.

9.3. Matrix Groups

Now we shall consider a group $\mathbf{G}$ which is a smooth subgroup of the group $\mathbf{GL}(m)$ of $m \times m$ nonsingular matrices. (The notation stems from the common name, *general linear group*.)

Suppose $u^1, \cdots, u^n$ is a coordinate system on $\mathbf{G}$ in some neighborhood of I, the identity matrix, and that $X = X(u^1, \cdots, u^n)$ is a typical point in this neighborhood. The matrix dX of one-forms certainly contains n linearly independent one-forms because the n-dimensional group $\mathbf{G}$ is smoothly imbedded in $\mathbf{GL}(m)$. Consequently the matrix

$$\Omega = X^{-1}\, dX$$

of one-forms contains n linearly independent ones. But *each element of* Ω *is left invariant.* For if A is any fixed element of $\mathbf{G}$, the left translation by A is given by

$$X \longrightarrow AX,$$

while

$$(AX)^{-1}\, d(AX) = (X^{-1}A^{-1})(A\, dX) = X^{-1}\, dX.$$

This allows us to make explicit calculations in many important groups.

Next we note an important geometric interpretation of Ω. We interpret each element X of $\mathbf{G}$ as a linear transformation on the space $\mathbf{E}^n$ of row vectors $\mathbf{v} = (v_1, \cdots, v_n)$. Thus

$$\mathbf{v} \longrightarrow \mathbf{w} = \mathbf{v}X.$$

We ask, how does $d\mathbf{w}$ grow out of $\mathbf{w}$ under the group action? Here $\mathbf{v}$ is fixed and X varies over $\mathbf{G}$. We have

$$d\mathbf{w} = \mathbf{v}\, dX = (\mathbf{w}X^{-1})\, dX,$$

$$d\mathbf{w} = \mathbf{w}\Omega.$$

This means Ω can be interpreted as an "infinitesimal group element." (Cf. Section 4.2.)

One final remark. The constants of structure can often be explicitly obtained from these considerations:

$$\Omega = X^{-1}dX$$

$$dX = X\Omega,$$

$$0 = d(dX) = dX\,\Omega + X\,d\Omega$$

$$= (X\Omega)\,\Omega + X\,d\Omega.$$

Hence,

$$d\Omega + \Omega^2 = 0.$$

9.4. Examples of Matrix Groups

Example 3.

$$\mathbf{G} = \left\{\begin{pmatrix} x & y \\ 0 & 1 \end{pmatrix} \mid x > 0\right\},$$

the proper affine group on the line. One easily sees the isomorphism between $\mathbf{G}$ and the transformation group

$$t \longrightarrow xt + y.$$

Here

$$X = \begin{pmatrix} x & y \\ 0 & 1 \end{pmatrix}, \qquad X^{-1} = \frac{1}{x}\begin{pmatrix} 1 & -y \\ 0 & x \end{pmatrix},$$

$$\Omega = \frac{1}{x}\begin{pmatrix} 1 & -y \\ 0 & x \end{pmatrix}\begin{pmatrix} dx & dy \\ 0 & 0 \end{pmatrix} = \frac{1}{x}\begin{pmatrix} dx & dy \\ 0 & 0 \end{pmatrix}.$$

Hence $\sigma^1 = dx/x$, $\sigma^2 = dy/x$ are left invariant. The left invariant volume element is

$$\sigma^1 \wedge \sigma^2 = \frac{dx\,dy}{x^2}.$$

Since $d\sigma^1 = 0$, $d\sigma^2 = -dx\,dy/x^2 = -\sigma^1 \wedge \sigma^2$, the only significant constant of structure is

$$c_{12}^2 = -c_{21}^2 = -1.$$

If we seek *right invariant* forms, we find them in

$$(dX)X^{-1} = \frac{1}{x}\begin{pmatrix} dx & dy \\ 0 & 0 \end{pmatrix}\begin{pmatrix} 1 & -y \\ 0 & x \end{pmatrix} = \frac{1}{x}\begin{pmatrix} dx & -y\,dx + x\,dy \\ 0 & 0 \end{pmatrix},$$

so a basis is

$$\alpha^1 = \sigma^1 = \frac{dx}{x}, \qquad \alpha^2 = \frac{-y\,dx + x\,dy}{x}.$$

The right invariant volume element is

$$\alpha^1 \wedge \alpha^2 = \frac{dx\,dy}{x},$$

very different from the left invariant one. Also

$$d\alpha^2 = \frac{dx\,dy}{x} = \alpha^1 \wedge \alpha^2.$$

We shall compute the effect on σ^2 of the right translation R_A, where

$$A = \begin{pmatrix} a & b \\ 0 & 1 \end{pmatrix}.$$

We have

$$R_A(X) = XA = \begin{pmatrix} x & y \\ 0 & 1 \end{pmatrix}\begin{pmatrix} a & b \\ 0 & 1 \end{pmatrix} = \begin{pmatrix} ax & bx + y \\ 0 & 1 \end{pmatrix},$$

$$R_A{}^*\sigma^2 = \frac{d(bx + y)}{ax} = \frac{b}{a}\frac{dx}{x} + \frac{1}{a}\frac{dy}{x} = \frac{b}{a}\sigma^1 + \frac{1}{a}\sigma^2.$$

Example 4. The *step transformation group* of all matrices

$$X = \begin{pmatrix} x & y \\ 0 & x \end{pmatrix}, \qquad x > 0.$$

$$\Omega = X^{-1}\,dX = \frac{1}{x^2}\begin{pmatrix} x & -y \\ 0 & x \end{pmatrix}\begin{pmatrix} dx & dy \\ 0 & dx \end{pmatrix} = \frac{1}{x^2}\begin{pmatrix} x\,dx & x\,dy - y\,dx \\ 0 & x\,dx \end{pmatrix}.$$

We may take

$$\sigma^1 = \frac{dx}{x} = d(\ln x), \qquad \sigma^2 = \frac{x\,dy - y\,dx}{x^2} = d\left(\frac{y}{x}\right).$$

The choice of new coordinates

$$u = \ln x, \qquad v = \frac{y}{x}$$

follows the procedure in Example 2.

Thus

$$X = \begin{pmatrix} e^u & ve^u \\ 0 & e^u \end{pmatrix} = e^u\begin{pmatrix} 1 & v \\ 0 & 1 \end{pmatrix}.$$

If

$$A = e^a\begin{pmatrix} 1 & b \\ 0 & 1 \end{pmatrix}$$

is another, then

$$AX = e^{a+u}\begin{pmatrix} 1 & b+v \\ 0 & 1 \end{pmatrix}.$$

The invariant volume element is $dx\,dy/x^2 = du\,dv$.

Example 5. $\mathbf{G} = \mathbf{GL}(n)$, the *general linear group* of all nonsingular $n \times n$ matrices. The general element is $X = \|x_i{}^j\|$ where the $x_i{}^j$ are independent variables (subject to the inequality $\Delta = \det(x_i{}^j) \neq 0$). Set

$$Y = \operatorname{cof} X, \qquad X^{-1} = \frac{1}{\Delta} Y.$$

We have

$$\Omega = X^{-1}\,dX = \|\sigma_i{}^k\|,$$

where

$$\sigma_i{}^k = \frac{1}{\Delta} \sum y_i{}^j\,dx_j{}^k.$$

The n^2 left invariant forms $\sigma_i{}^k$ are necessarily linearly independent. We compute the volume element

$$\tau = \prod_{i,k} \sigma_i{}^k$$

in two steps:

$$\tau = \nu^1 \wedge \cdots \wedge \nu^n,$$

$$\nu^k = \sigma_1{}^k \wedge \cdots \wedge \sigma_n{}^k = \frac{1}{\Delta^n} \left(\sum y_1{}^j\,dx_j{}^k\right) \wedge \cdots \wedge \left(\sum y_n{}^j\,dx_j{}^k\right)$$

$$= \frac{1}{\Delta^n} \det(y_i{}^j)\,dx_1{}^k \cdots dx_n{}^k.$$

From $XY = \Delta I$ we have $\det(X)\det(Y) = \Delta^n$, hence

$$\det(Y) = \Delta^{n-1},$$

$$\nu^k = \frac{1}{\Delta}\,dx_1{}^k \cdots dx_n{}^k,$$

$$\tau = \frac{1}{\Delta^n} \prod_{k=1}^{n} (dx_1{}^k \cdots dx_n{}^k).$$

It is clear from this that the right invariant volume element will also be τ, which is an unusual feature of this group since it is highly non-abelian.

Example 6. $\mathbf{G} = \mathbf{SL}(n)$, the *special linear* or *unimodular group* of all $n \times n$ matrices of determinant one. The special feature we shall note is

$$\operatorname{trace}(\Omega) = 0.$$

This follows from a general formula for a matrix function X of any number of variables. Set $\Delta = \det(X)$. The formula is

$$\frac{d\Delta}{\Delta} = \operatorname{trace}(X^{-1}\,dX).$$

This is proved as follows. Denote by

$$\mathbf{c}^j = \begin{bmatrix} x_1{}^j \\ x_2{}^j \\ \vdots \\ x_n{}^j \end{bmatrix}$$

the jth column of X and consider

$$\Delta = \det(\mathbf{c}^1, \cdots, \mathbf{c}^n)$$

as a function of the columns. Then

$$\begin{aligned} d\Delta &= \sum_{j=1}^{n} \Delta(\mathbf{c}^1, \cdots, \mathbf{c}^{j-1}, d\mathbf{c}^j, \mathbf{c}^{j+1}, \cdots, \mathbf{c}^n) \\ &= \sum_{j=1}^{n}\sum_{i=1}^{n} y_j{}^i\,dx_i{}^j = \operatorname{trace}[(\operatorname{cof} X)\,dX] \\ &= \operatorname{trace}[\Delta X^{-1}\,dX] = \Delta\operatorname{trace}(X^{-1}\,dX). \end{aligned}$$

For $\mathbf{G} = \mathbf{SL}(n)$ we have $\Delta = 1$, $d\Delta = 0$, hence $\operatorname{trace}(\Omega) = 0$. For $n = 2$ we have

$$X = \begin{pmatrix} x & y \\ u & v \end{pmatrix}, \qquad \Delta = xv - yu = 1,$$

$$\begin{aligned} \Omega = X^{-1}\,dX &= \begin{pmatrix} v & -y \\ -u & x \end{pmatrix}\begin{pmatrix} dx & dy \\ du & dv \end{pmatrix} \\ &= \begin{pmatrix} v\,dx - y\,du & v\,dy - y\,dv \\ -u\,dx + x\,du & -u\,dy + x\,dv \end{pmatrix} = \begin{pmatrix} \sigma^1 & \sigma^2 \\ \sigma^3 & \sigma^4 \end{pmatrix}. \end{aligned}$$

Differentiating $\Delta = 1$ yields

$$x\,dv + v\,dx - y\,du - u\,dy = 0, \qquad \sigma^1 + \sigma^4 = 0.$$

For the left invariant volume element we may take

$$\tau = \sigma^1 \wedge \sigma^3 \wedge \sigma^2 = dx\,du(v\,dy - y\,dv)$$
$$= v\,dx\,du\,dy - y\,dx\,du\,dv.$$

Example 7. $\mathbf{G} = \mathbf{0}^+(n)$, the proper orthogonal group of all $n \times n$ matrices X for which

$${}^tX = X^{-1}, \qquad \det(X) = +1.$$

Here the superscript t denotes *transpose*. The essential feature about Ω is that it is a skew-symmetric matrix,

$$\Omega + {}^t\Omega = 0.$$

Because the group $\mathbf{G}$ has dimension $n(n-1)/2$, it follows that the elements above the main diagonal in Ω form a basis for left invariant one-forms and their product is the left invariant volume element. We establish this property of Ω as follows:

$$X\,{}^tX = I, \qquad (dX)\,{}^tX + X\,{}^t(dX) = 0,$$
$$X^{-1}\,dX + {}^t(dX)\,{}^tX^{-1} = 0,$$
$$X^{-1}\,dX + {}^t(X^{-1}\,dX) = 0, \qquad \Omega + {}^t\Omega = 0.$$

For $n = 2$,

$$X = \begin{pmatrix} \cos\theta & \sin\theta \\ -\sin\theta & \cos\theta \end{pmatrix},$$

$$\Omega = X^{-1}\,dX = \begin{pmatrix} \cos\theta & -\sin\theta \\ \sin\theta & \cos\theta \end{pmatrix}\begin{pmatrix} -\sin\theta & \cos\theta \\ -\cos\theta & -\sin\theta \end{pmatrix} d\theta = \begin{pmatrix} 0 & 1 \\ -1 & 0 \end{pmatrix} d\theta.$$

For $n > 2$ the calculation becomes complicated and hinges on explicit parametrizations of $\mathbf{G}$. The cases of even and odd n are rather different.

9.5. Bi-invariant Forms

We take a Lie group $\mathbf{G}$ with identity element e. Because $\mathbf{G}$ is a manifold, there is a coordinate neighborhood $\mathbf{U}$ of e with a local coordinate system such that the coordinates of e are $(0, 0, \cdots, 0)$. Suppose x and y are very near to e and in $\mathbf{U}$. Then $z = xy$ is in $\mathbf{U}$. If the coordinates of these three points are $(x^1, \cdots, x^n)$, $(y^1, \cdots, y^n)$, $(z^1, \cdots, z^n)$, respectively, we may write

$$z^i = z^i(x^1, \cdots, x^n, y^1, \cdots, y^n).$$

Since $xe = x$ and $ey = y$, we have

$$z^i(x^1, \cdots, x^n, 0, \cdots, 0) = x^i,$$
$$z^i(0, \cdots, 0, y^1, \cdots, y^n) = y^i,$$

and because of these facts,

$$z^i = x^i + y^i + (\text{higher order terms in the } x^j \text{ and } y^k).$$

In particular, if $y = x^{-1}$, then $z = e$, and

$$0 = x^i + y^i + (\text{higher order terms}).$$

We apply these simple remarks as follows. Let ψ denote the mapping $\psi(x) = x^{-1}$,

$$\psi: \quad \mathbf{G} \longrightarrow \mathbf{G}.$$

If we write

$$y = \psi(x),$$
$$y^i = y^i(x^1, \cdots, x^n),$$

then by the relation just discussed,

$$\left.\frac{\partial y^i}{\partial x^j}\right|_0 = -\delta^i_j.$$

This means that

$$y^i = -x^i + (\text{higher order terms in the } x^j).$$

We may also express this another way:

$$\psi: \quad (x^1, \cdots, x^n) \longrightarrow -(x^1, \cdots, x^n) + (\text{higher order terms}).$$

Since $\psi(e) = e^{-1} = e$, the induced mapping ψ^* takes each differential form at e to another form at e. Evidently we have

$$\psi^*(dx^i) = -dx^i \qquad (\text{at } e),$$
$$\psi^*(dx^1 \cdots dx^p) = (-1)^p(dx^1 \cdots dx^p) \qquad (\text{at } e).$$

Thus *if ω_e is any p-form at e, then*

$$\psi^*(\omega_e) = (-1)^p \omega_e.$$

For each y in $\mathbf{G}$, the right translation R_y was defined by

$$R_y(x) = xy.$$

A form ω is called *right invariant* if

$$R_y{}^* \omega = \omega$$

for each y in $\mathbf{G}$. Our first application of the map ψ is the following:

A form ω is left invariant if and only if $\psi^* \omega$ is right invariant.

For

$$\psi(R_y(x)) = \psi(xy) = (xy)^{-1} = y^{-1}x^{-1}$$
$$= L_{y^{-1}}(x^{-1}) = L_{y^{-1}}(\psi(x)),$$

hence

$$\psi \circ R_y = L_{y^{-1}} \circ \psi,$$
$$R_y^* \circ \psi^* = \psi^* \circ L_{y^{-1}}^*.$$

If ω is left invariant, then for each y in **G**

$$R_y^*(\psi^* \omega) = \psi^*(L_{y^{-1}}^* \omega) = \psi^* \omega,$$

hence $\psi^* \omega$ is right invariant. Similarly, if ω is right invariant then $\psi^* \omega$ is left invariant. Since $(x^{-1})^{-1} = x$, $\psi \circ \psi = \iota$ and so $\psi^*(\psi^* \omega) = \omega$. It follows that if $\psi^* \omega$ is right invariant then ω is left invariant.

Next we shall see that using right invariant forms instead of left invariant ones does not give additional constants of structure. For let $\sigma^1, \cdots, \sigma^n$ be a basis of the left invariant one-forms. Then the corresponding constants of structure are read from the equations

$$d\sigma^i = \tfrac{1}{2} \sum c^i{}_{jk} \sigma^j \wedge \sigma^k.$$

The forms $\tau^1 = \psi^* \sigma^1, \cdots, \tau^n = \psi^* \sigma^n$ are now a basis for the right invariant one-forms. But ψ^* applied to our equation yields, since the c's are constants,

$$\psi^*(d\sigma^i) = \tfrac{1}{2} \sum c^i{}_{jk} \psi^*(\sigma^j \wedge \sigma^k),$$
$$d(\psi^* \sigma^i) = \tfrac{1}{2} \sum c^i{}_{jk} (\psi^* \sigma^j) \wedge (\psi^* \sigma^k),$$
$$d\tau^i = \tfrac{1}{2} \sum c^i{}_{jk} \tau^j \wedge \tau^k.$$

Now we pass on to the study of *bi-invariant forms*, i.e., forms which are both left and right invariant. We derive one important result.

Let ω be a bi-invariant p-form. Then

$$d\omega = 0.$$

For $\psi^* \omega$ is left invariant since ω is right invariant. We know from our calculation on the previous page that at the point e,

$$\psi^*(\omega_e) = (-1)^p \omega_e.$$

But ω and $\psi^*(\omega)$ are both left invariant, hence what is true at e is true everywhere,

$$\psi^*(\omega) = (-1)^p \omega.$$

On the other hand, $d\omega$ is a $(p+1)$-form, also bi-invariant, so the same conclusion applies:

$$\psi^*(d\omega) = (-1)^{p+1} d\omega.$$

But,

$$\psi^*(d\omega) = d(\psi^* \omega) = d[(-1)^p \omega] = (-1)^p d\omega.$$

From these equations follows $d\omega = 0$.

We apply this to the case in which $\mathbf{G}$ is a commutative group. Then the left and right translations are the same thing so that each left invariant form is bi-invariant. In particular if $\sigma^1, \cdots, \sigma^n$ is a basis of left invariant one-forms, each $d\sigma^i = 0$; the constants of structure all vanish. In Example 2 of Section 9.2 we showed that any group with vanishing structure constants has the local structure of Euclidean space (and incidentally is commutative).

Here is one more result on bi-invariant forms which goes in a different direction.

Let $\mathbf{G}$ be an n-dimensional closed (compact) Lie group and let ω be a left invariant n-form on $\mathbf{G}$. Then ω is bi-invariant.

For each x in $\mathbf{G}$, $R_x{}^*\omega$ is also left invariant. Assuming $\omega \neq 0$, we have $R_x{}^*\omega = f(x)\omega$, where $f(x)$ is a real number, since the space of left invariant n-forms has dimension one. Because $R_x{}^* \circ R_y{}^* = R_{yx}{}^*$, we have $f(xy) = f(x)f(y)$. (Real numbers commute!) Now $f(x)$ never vanishes since $1 = f(e) = f(x)f(x^{-1})$. Thus f maps $\mathbf{G}$ into the reals $\mathbf{R}$ with 0 removed. Since $\mathbf{G}$ is compact the image of $\mathbf{G}$ under f is a bounded interval in $\mathbf{R}$, bounded away from zero. It is also a subgroup of the multiplicative group of positive reals since f preserves multiplication. [Positive because the image $f(\mathbf{G})$ of the manifold is an interval and it contains $1 = f(e)$.] If $f(\mathbf{G})$ contains any real number $a \neq 1$, then $a^n \longrightarrow \infty$ or $a^n \longrightarrow 0$, both impossible, since a^n must remain in the interval $f(\mathbf{G})$ which is closed under multiplication. Hence $f(\mathbf{G})$ consists of 1 alone, $f(x) = 1$ for each x in $\mathbf{G}$, $R_x{}^*\omega = \omega$, ω is bi-invariant as asserted. (We have used the fact that $\mathbf{G}$ is a manifold, hence connected, to conclude that $f(\mathbf{G})$ is an interval.)

9.6. Problems

1. Let $\mathbf{C}^*$ denote the multiplicative group of nonzero complex numbers. Find the invariant volume (area) element.

2. Consider the 4-dimensional group of all matrices

$$\begin{pmatrix} z & w \\ 0 & 1 \end{pmatrix}$$

where z and w are complex numbers, $z \neq 0$. Determine constants of structure. Show that the left invariant volume element is

$$-(dz \wedge d\bar{z} \wedge dw \wedge d\bar{w})/|z|^4.$$

Here $d\bar{z} = dx - i\,dy$ if $dz = dx + i\,dy$.

3. Discuss other groups of complex matrices analogous to the examples of Section 9.3. For example, discuss the relation between unitary matrices and skew-hermitian ones.

4. Extend the coordinate considerations of Section 9.5 by showing that

$$z^i = x^i + y^i + \sum a^i{}_{jk} x^j y^k + (\text{terms of order three and higher}).$$

Show also that

$$(a^i{}_{jk} - a^i{}_{kj})$$

are the constants of structure for a suitable basis of left invariant one-forms. Compare xy and yx.

5. We know that each left invariant p-form can be expressed in terms of left invariant one-forms. Does a corresponding result hold for bi-invariant forms?

6. Let $c^i{}_{jk}$ be constants of structure of a group and set

$$g_{jl} = \sum_{i,k} c^i{}_{jk} c^k{}_{il}.$$

Show that g_{jl} is a symmetric tensor. Now set

$$c_{ijk} = \sum_{l} c^l{}_{ij} g_{lk}.$$

Show that c_{ijk} is a skew-symmetric tensor.

7. Let $\mathbf{z}$ be a p-cycle on $\mathbf{G}$. Show that for each closed p-form ω and each g in $\mathbf{G}$,

$$\int_{\mathbf{z}} \omega = \int_{\mathbf{z}} L_g{}^* \omega.$$

(We must assume that $\mathbf{G}$ is connected, i.e., consists of one piece only.)

X

Applications to Physics

10.1. Phase and State Space

We propose to study a holonomic mechanical system with a finite number of degrees of freedom, avoiding collision phenomena. In this section we formulate the geometry of such a system.

The *position space* is simply an n-dimensional manifold **M**.

We next define the *phase space* attached to **M**. This is the space of all covariant vectors at all points of **M**. To make this precise, we consider a coordinate patch **U** on **M** with local coordinates

$$q^1, \cdots, q^n.$$

At a point P of **U**, a covariant vector is simply a one-form at P, hence is given by its components

$$p_1, \cdots, p_n, \qquad p_i \text{ real}$$

(where the one-form itself is $\sum p_i dq^i$).

If

$$\bar{q}^1, \cdots, \bar{q}^n$$

is another local coordinate system valid at P, then the components of the same covariant vector with respect to the $\bar{q}^i$ are

$$\bar{p}_i = \sum p_j \frac{\partial q^j}{\partial \bar{q}^i}.$$

The totality of all such covariant vectors at all points of **M** constitutes the $(2n)$-dimensional *phase space* **P**. To each coordinate neighborhood **U** on **M** with local coordinates $q^1, \cdots, q^n$ corresponds the coordinate neighborhood $\mathbf{U} \times \mathbf{E}^n$ with local coordinates

$$q^1, \cdots, q^n, p_1, \cdots, p_n.$$

It follows that the one-form

$$\alpha = \sum p_i dq^i$$

is a one-form on **P**, entirely independent of local coordinates. We have

$$d\alpha = \sum dp_i dq^i$$

so that the *phase density* (see Section 2.3)

$$dp_1 \cdots dp_n dq^1 \cdots dq^n$$

is a $2n$-form on **P**, never zero, defined by

$$\pm (d\alpha)^n = (n!)(dp_1 \cdots dp_n dq^1 \cdots dq^n),$$

and serves us as a volume element on **P**.

We shall derive some useful relations from the transformation of coordinates

$$\begin{cases} \bar{q}^i = \bar{q}^i(q^1, \cdots, q^n) \\ \bar{p}_i = \sum p_j \dfrac{\partial q^j}{\partial \bar{q}^i} \end{cases} \qquad (i = 1, \cdots, n)$$

valid on the overlap of local coordinate neighborhoods **U** and $\bar{\mathbf{U}}$.

We set

$$\mathbf{q} = (q^1, \cdots, q^n) \quad \text{and} \quad \mathbf{p} = \begin{pmatrix} p_1 \\ \vdots \\ p_n \end{pmatrix}$$

and define $\bar{\mathbf{q}}$ and $\bar{\mathbf{p}}$ similarly. Then $\bar{\mathbf{q}} = \bar{\mathbf{q}}(\mathbf{q})$ implies

$$d\bar{\mathbf{q}} = d\mathbf{q}A, \quad A = \left(\frac{\partial \bar{q}^j}{\partial q^i}\right) = A(\mathbf{q}).$$

Since $\alpha = d\mathbf{q}\,\mathbf{p} = d\bar{\mathbf{q}}\,\bar{\mathbf{p}}$, we have

$$\mathbf{p} = A\bar{\mathbf{p}}, \text{ that is, } \bar{\mathbf{p}} = \bar{A}\mathbf{p},$$

where $\bar{A} = A^{-1}$ is also a Jacobean matrix. From

$$d\bar{\mathbf{p}} = d\bar{A}\mathbf{p} + \bar{A}d\mathbf{p} \quad \text{and} \quad d\mathbf{q} = d\bar{\mathbf{q}}\bar{A}$$

we deduce first that $(\partial \bar{p}_i/\partial p_j) = \bar{A} = (\partial q^j/\partial \bar{q}^i)$.

To continue, we note two relations. From $\bar{A} = A^{-1}$ we have

$$d\bar{A} = -A^{-1}\,dA\,A^{-1} = -\bar{A}\,dA\,\bar{A}.$$

From $d\bar{\mathbf{q}} = d\mathbf{q}A$ we have $0 = d\mathbf{q}\,dA$, which easily implies

$$\frac{\partial a^l{}_k}{\partial q^j} = \frac{\partial a^l{}_j}{\partial q^k}.$$

Now

$$\begin{aligned} d\bar{\mathbf{p}} = d\bar{A}\mathbf{p} + \bar{A}d\mathbf{p} &= -\bar{A}dA\bar{A}\,\mathbf{p} + \bar{A}d\mathbf{p} \\ &= -\bar{A}dA\bar{\mathbf{p}} + \bar{A}d\mathbf{p}. \end{aligned}$$

Also

$$d\mathbf{p} = d(A\bar{\mathbf{p}}) = dA\bar{\mathbf{p}} + Ad\bar{\mathbf{p}}.$$

Thus $\partial \bar{p}_i/\partial q^k$ is the coefficient of dq^k in the i-th row of $-\bar{A}\,dA\,\bar{\mathbf{p}}$ and $\partial p_k/\partial \bar{q}^i$ is the coefficient of $d\bar{q}^i$ in the k-th row of $dA\,\bar{\mathbf{p}}$. Now

$$\bar{A}\,dA = \left(\sum \bar{a}^j{}_i \frac{\partial a^l{}_j}{\partial q^k}\,\mathrm{d}q^k\right)$$

and

$$dA = \left(\sum \frac{\partial a^l{}_k}{\partial q^i}\,\mathrm{d}\bar{q}^i\right) = \left(\sum \bar{a}^j{}_i \frac{\partial a^l{}_k}{\partial q^j}\,\mathrm{d}\bar{q}^i\right)$$
$$= \left(\sum \bar{a}^j{}_i \frac{\partial a^l{}_j}{\partial q^k}\,\mathrm{d}\bar{q}^i\right).$$

Therefore the coefficient of dq^k in the i-th row of $-\bar{A}d\bar{A}\bar{\mathbf{p}}$ and the coefficient of $d\bar{q}^i$ in the k-th row of $dA\bar{\mathbf{p}}$ are respectively

$$\frac{\partial \bar{p}_i}{\partial q^k} = -\sum \bar{a}^j{}_i \frac{\partial a^l{}_j}{\partial q^k}\bar{p}_l \quad \text{and} \quad \frac{\partial p_k}{\partial q_i} = \sum \bar{a}^j{}_i \frac{\partial a^l{}_j}{\partial q^k}\bar{p}_l\,.$$

We conclude that $\partial\bar{p}_i/\partial q^k = -\partial p^k/\partial\bar{q}^i$, so we have proved

$$\begin{cases} \dfrac{\partial \bar{p}_i}{\partial q^k} = -\dfrac{\partial p_k}{\partial \bar{q}^i} \\[2ex] \dfrac{\partial \bar{p}_i}{\partial p_j} = \dfrac{\partial q^j}{\partial \bar{q}^i}\,. \end{cases}$$

Finally, the *state space* is the product

$$\mathbf{S} = \mathbf{P} \times \mathbf{E}^1$$

a $(2n + 1)$ dimensional space. We think of $\mathbf{E}^1$ as the time axis. Local coordinates for $\mathbf{S}$ are

$$q^1, \cdots, q^n, \quad p_1, \cdots, p_n, \quad t.$$

10.2. Hamiltonian Systems

We wish to consider a dynamical system in Hamiltonian form. We begin by tracing the evolution of this from Lagrange's equations of motion, which in Euclidean coordinates reduce to Newton's law of motion. We deal only with conservative holonomic systems.

The treatment first of all is local. We deal with a coordinate patch in $q^1, \cdots, q^n$ space. For each instant of time, there is a point (position)

$$(q^1(t), \cdots, q^n(t))$$

which represents the trajectory of the system. As is customary, we set $\dot{u} = du/dt$. The *kinetic energy* is a function

$$T(q^1, \cdots, q^n, \dot{q}^1, \cdots, \dot{q}^n)$$

which is supposed to be a positive definite quadratic form in the variables $\dot{q}^i$. The *potential energy* is a function

$$V = V(q^1, \cdots, q^n, t)$$

and the *Lagrangian function*, or *kinetic potential*, is

$$L = T - V.$$

The differential equations of motion are then

$$\frac{d}{dt}\left(\frac{\partial L}{\partial \dot{q}^i}\right) - \frac{\partial L}{\partial q^i} = 0 \qquad (i = 1, \cdots, n).$$

For the first term we have

$$\frac{d}{dt}\left(\frac{\partial L}{\partial \dot{q}^i}\right) = \frac{\partial^2 L}{\partial \dot{q}^i\, \partial t} + \frac{\partial^2 L}{\partial \dot{q}^i\, \partial q^j}\dot{q}^j + \frac{\partial^2 L}{\partial \dot{q}^i\, \partial \dot{q}^k}\ddot{q}^k$$

so that the Lagrange equations are a system of n second order ordinary equations for the unknowns $q^1, \cdots, q^n$. We now convert these to a system of $2n$ first order equations in $2n$ unknowns.

We introduce the *generalized momentum components*

$$p_1, \cdots, p_n$$

by

$$p_i = \frac{\partial L}{\partial \dot{q}^i} = \frac{\partial T}{\partial \dot{q}^i}.$$

Because the quadratic form T is definite, the transformation of variables

$$(q^1, \cdots, q^n, \dot{q}^1, \cdots, \dot{q}^n) \longleftrightarrow (q^1, \cdots, q^n, p_1, \cdots, p_n)$$

is a smooth one both ways.

To reach the Hamilton form, we shall follow tradition and use a rather confusing notation. The matter was better expressed in Section 3.5.

The function T is always considered as a function of the $2n$ variables

$$q^1, \cdots, q^n, \dot{q}^1, \cdots, \dot{q}^n.$$

The function V which involves the q^i (and t) alone may be considered as a function on the space of variables $q^1, \cdots, q^n, \dot{q}^1, \cdots, \dot{q}^n, t$ or on the space of variables $q^1, \cdots, q^n, p_1, \cdots, p_n, t$.

We introduce the *Hamiltonian*

$$H = H(q^1, \cdots, q^n, p_1, \cdots, p_n, t) = \sum p_i\dot{q}^i - L,$$

always considered as a function on the space of variables

$$q^1, \cdots, q^n, p_1, \cdots, p_n, t.$$

Since T is homogeneous quadratic in the $\dot{q}^i$ we have

$$2T = \sum \dot{q}^i \frac{\partial T}{\partial \dot{q}^i} = \sum p_i \dot{q}^i,$$

hence $H = 2T - L = 2T - (T - V)$,

$$H = T + V,$$

and so H represents *total energy*.

From

$$2T = \sum p_i \dot{q}^i$$

we have

$$2\,dT = \sum p_i \,d\dot{q}^i + \sum \dot{q}^i \,dp_i.$$

But

$$\begin{aligned} dT &= \sum \left(\frac{\partial T}{\partial q^i}\,dq^i + \frac{\partial T}{\partial \dot{q}^i}\,d\dot{q}^i\right) \\ &= \sum \frac{\partial T}{\partial q^i}\,dq^i + \sum p_i \,d\dot{q}^i. \end{aligned}$$

Subtracting,

$$dT = -\sum \frac{\partial T}{\partial q^i}\,dq^i + \sum \dot{q}^i \,dp_i.$$

From this,

$$\begin{aligned} dH = dT + dV &= \sum \left(-\frac{\partial T}{\partial q^i} + \frac{\partial V}{\partial q^i}\right) dq^i + \sum \dot{q}^i \,dp_i \\ &= -\sum \frac{\partial L}{\partial q^i}\,dq^i + \sum \dot{q}^i \,dp_i. \end{aligned}$$

Thus

$$\begin{cases} \dfrac{\partial H}{\partial q^i} = -\dfrac{\partial L}{\partial q^i} = -\dfrac{d}{dt}\left(\dfrac{\partial L}{\partial \dot{q}^i}\right) = -\dot{p}_i, \\[2ex] \dfrac{\partial H}{\partial p_i} = \dot{q}^i. \end{cases}$$

This gives us the *equations of motion in Hamilton, or canonical*, form:

$$\begin{cases} \dot{q}^i = \dfrac{\partial H}{\partial p_i} \\[2ex] \dot{p}_i = -\dfrac{\partial H}{\partial q^i} \end{cases} \qquad (i = 1, \cdots, n).$$

We next check what functions H qualify as Hamiltonians. We write

$$T = \tfrac{1}{2} \sum a_{ij}(q)\dot{q}^i \dot{q}^j$$

where $\|a_{ij}(q)\|$ is a symmetric positive definite matrix function of the position variables q. It is convenient to set

$$\|b^{ij}(q)\| = \|a_{ij}(q)\|^{-1},$$

also symmetric, positive definite. Then

$$p_i = \frac{\partial T}{\partial \dot{q}^i} = \sum a_{ij}\dot{q}^j,$$

which we invert to

$$\dot{q}^i = \sum b^{ik} p_k .$$

This gives us

$$\begin{aligned} T &= \tfrac{1}{2} \sum a_{ij} b^{ik} b^{jl} p_k p_l \\ &= \tfrac{1}{2} \sum b^{jl} p_j p_l . \end{aligned}$$

From this,

$$H(q, p, t) = \tfrac{1}{2} \sum b^{ij}(q) p_i p_j + V(q, t).$$

This shows us the form of any Hamiltonian function.

We now wish to formulate Hamiltonian mechanics globally. To discover the correct approach, we compare two Hamiltonian systems

$$\begin{bmatrix} H = T - V \\ T = \tfrac{1}{2} \sum b^{ij}(q) p_i p_j \\ V = V(q, t) \\ \dot{q}^i = \dfrac{\partial H}{\partial p_i} \\ \dot{p}_i = -\dfrac{\partial H}{\partial q^i} \end{bmatrix} \qquad \begin{bmatrix} \bar{H} = \bar{T} - \bar{V} \\ \bar{T} = \tfrac{1}{2} \sum \bar{b}^{ij}(\bar{q}) \bar{p}_i \bar{p}_j \\ \bar{V} = \bar{V}(\bar{q}, t) \\ \dot{\bar{q}}^i = \dfrac{\partial \bar{H}}{\partial \bar{p}_i} \\ \dot{\bar{p}}_i = -\dfrac{\partial \bar{H}}{\partial \bar{q}^i} \end{bmatrix}$$

which are supposed to be defined on intersecting regions $\mathbf{U}$, $\bar{\mathbf{U}}$ and be equivalent on the common part $\mathbf{U} \cap \bar{\mathbf{U}}$ of $\mathbf{U}$ and $\bar{\mathbf{U}}$. The coordinate transformation

$$(q, p, t) \longleftrightarrow (\bar{q}, \bar{p}, t)$$

is given as in Section 10.1:

$$\begin{cases} \bar{q}^i = \bar{q}^i(q^1, \cdots, q^n) \\ \bar{p}_i = \sum p_j \dfrac{\partial q^j}{\partial \bar{q}^i} . \end{cases}$$

We have

$$\sum \bar{b}^{ij}\bar{p}_j = \frac{\partial \bar{H}}{\partial \bar{p}_i} = \dot{\bar{q}}^i = \sum \frac{\partial \bar{q}^i}{\partial q^k}\dot{q}^k$$
$$= \sum \frac{\partial \bar{q}^i}{\partial q^k}\frac{\partial H}{\partial p_k} = \sum \frac{\partial \bar{q}^i}{\partial q^k} b^{kl} p_l .$$

Hence

$$\bar{T} = \tfrac{1}{2}\sum \bar{p}_i \bar{b}^{ij}\bar{p}_j = \tfrac{1}{2}\sum \bar{p}_i \frac{\partial \bar{q}^i}{\partial q^k} b^{kl} p_l$$
$$= \tfrac{1}{2}\sum p_k b^{kl} p_l = T,$$

and so

$$\bar{T} = T$$

on the common part of **U** and $\bar{\mathbf{U}}$, hence also

$$\bar{H} - H = \bar{V} - V.$$

Using the symmetry relation,

$$\frac{\partial \bar{p}_i}{\partial q^k} + \frac{\partial p_k}{\partial \bar{q}^i} = 0,$$

derived in the last section, and the remaining equations of motion:

$$-\frac{\partial \bar{H}}{\partial \bar{q}^i} = \dot{\bar{p}}_i = \sum \frac{\partial \bar{p}_i}{\partial q^j}\dot{q}^j + \sum \frac{\partial \bar{p}_i}{\partial p_k}\dot{p}_k$$
$$= -\sum \frac{\partial p_j}{\partial \bar{q}^i}\frac{\partial H}{\partial p_j} - \sum \frac{\partial q^k}{\partial \bar{q}^i}\frac{\partial H}{\partial q^k}$$
$$= -\frac{\partial H}{\partial \bar{q}^i},$$
$$\frac{\partial \bar{H}}{\partial \bar{q}^i} = \frac{\partial H}{\partial \bar{q}^i}.$$

From this,

$$\frac{\partial}{\partial \bar{q}^i}(\bar{V} - V) = 0 \qquad (i = 1, \cdots, n).$$

It follows that $\bar{V} - V$ is a function of t alone,

$$\bar{V} = V + f(t),$$
$$\bar{H} = H + f(t),$$

this taking place on the intersection of **U** and $\bar{\mathbf{U}}$.

Having this, we can formulate what we mean by a global Hamiltonian system.

We begin with a position space **M** and form its derived spaces, phase space **P**, and state space **S**. We are give a function T on **P** such that over any local coordinate neighborhood **U** on **M** with coordinates $q^1, \cdots, q^n$ we have

$$T = \tfrac{1}{2} \sum b^{ij}(q) p_i p_j ,$$

a positive definite quadratic form. Here $(q^1, \cdots, q^n, p_1, \cdots, p_n)$ are the derived coordinates on the neighborhood $\mathbf{U} \times \mathbf{E}^n$ of **P** lying over **U**.

Now let $\mathbf{U}_\alpha, \mathbf{U}_\beta, \cdots$ denote the various local coordinate neighborhoods on **M**. For each one of these $\mathbf{U}_\alpha$ we have a function

$$V_\alpha = V_\alpha(q, t)$$

defined on

$$\mathbf{U}_\alpha \times \mathbf{E}^1.$$

Whenever $\mathbf{U}_\alpha$ overlaps $\mathbf{U}_\beta$, then on the common part $\mathbf{U}_\alpha \cap \mathbf{U}_\beta$ we have

$$V_\alpha - V_\beta = f_{\alpha\beta}(t),$$

a function of t alone. (Clearly $f_{\alpha\beta} + f_{\beta\alpha} = 0$, and

$$f_{\alpha\beta} + f_{\beta\gamma} + f_{\gamma\alpha} = 0 \qquad \text{on} \qquad \mathbf{U}_\alpha \cap \mathbf{U}_\beta \cap \mathbf{U}_\gamma .)$$

On the part of state space **S** lying over $\mathbf{U}_\alpha$, i.e., on $\mathbf{U}_\alpha \times \mathbf{E}^n \times \mathbf{E}^1$, set

$$H_\alpha = T + V_\alpha .$$

The *equations of motion* are given by

$$\begin{cases} \dot{q}^i = \dfrac{\partial H}{\partial p_i} \\ \dot{p}_i = -\dfrac{\partial H}{\partial q^i} \end{cases}$$

on $\mathbf{U}_\alpha \times \mathbf{E}^n \times \mathbf{E}^1$, where (q^i) are local coordinates on $\mathbf{U}_\alpha$. These are independent of the local coordinate systems (consistent) and define a motion (or flow) on all of **S**, always moving forward in the t, or time, direction.

Remark. It is customary in mechanics to exhibit the potential function so we have set down the functions V_α in our formulation. Actually the equations of motion merely require the gradient of the potential which is an intrinsic quantity on **S**. Precisely, there is a differential form ϖ on **S** so that locally

$$\varpi = \sum \frac{\partial V_\alpha}{\partial q^i} dq^i + \sum \frac{\partial V_\alpha}{\partial p_i} dp_i .$$

We can free ourselves altogether of reference to the functions V_α by requiring that there be given a one-form ϖ on $\mathbf{S}$ satisfying

(1) ϖ is free of dt,

(2) $$d\varpi \Big|_{dt=0} = 0.$$

By the converse of the Poincaré Lemma (Sections 3.6 and 3.7, especially the last remark on p. 31) this implies the existence of the functions V_α.

A *trajectory* of the motion is any particular solution of this system of differential equations. From the theory of ordinary differential equations we know that there is a unique trajectory through each point, so that state space $\mathbf{S}$ is smoothly filled with these curves. Along each one, t steadily increases, but not necessarily to arbitrarily large values. (For example, a particle may run off to infinity in finite time.)

Finally we note the *energy law*:

Along any trajectory,

$$\frac{dH}{dt} = \frac{\partial H}{\partial t}.$$

For

$$\begin{aligned} \frac{dH}{dt} = \dot{H} &= \sum \frac{\partial H}{\partial q^i} \dot{q}^i + \sum \frac{\partial H}{\partial p_i} \dot{p}_i + \frac{\partial H}{\partial t} \\ &= \sum (-\dot{p}_i \dot{q}^i) + \sum \dot{q}^i \dot{p}_i + \frac{\partial H}{\partial t} \\ &= \frac{\partial H}{\partial t}. \end{aligned}$$

Remark. Whether or not the functions $f_{\alpha\beta}(t)$ can be removed altogether by redefining each V_α so that there will be a single potential function V on all of $\mathbf{M} \times \mathbf{E}^1$ depends on two things, the manner in which the applied forces vary with time, and the topology of $\mathbf{M}$.

The topological difficulties are easily seen from the standard example of the steady magnetic field in the manifold $\mathbf{M}$, which consists of $\mathbf{E}^3$ minus the z-axis, due to a steady electric current in the z-axis. The trouble is that there are closed loops in $\mathbf{M}$ which are not boundaries of surfaces. (Cf. the situation in De Rham's theorems, Section 5.9.)

10.3. Integral-invariants

Over a local coordinate neighborhood $\mathbf{U} = \mathbf{U}_\alpha$ in position space $\mathbf{M}$ we consider the one-form

$$\omega = \omega_\alpha = \sum p_i \, dq^i - H \, dt.$$

This is defined on the portion $\mathbf{U} \times \mathbf{E}^n \times \mathbf{E}^1$ of state space which lies over $\mathbf{U}$. These forms ω_α do not necessarily fit together to make a one-form on $\mathbf{S}$ because on an intersection $\mathbf{U}_\alpha \cap \mathbf{U}_\beta$ we have

$$H_\alpha - H_\beta = V_\alpha - V_\beta = f_{\alpha\beta}(t).$$

If the V_α can be chosen so that all $f_{\alpha\beta}(t) = 0$, then we do have such a one-form on all of $\mathbf{S}$. This is exactly the case for a globally conservative system, the case in which the external forces are derived from a single potential function V. While this cannot be expected in general, we do see that

$$d\omega = d\omega_\alpha = \sum dp_i\, dq^i - dH\, dt$$

is a 2-form on all of $\mathbf{S}$, independent of local coordinates. This simply means that

$$d\omega_\alpha = d\omega_\beta$$

on $\mathbf{U}_\alpha \cap \mathbf{U}_\beta$, which is true because

$$dH_\beta\, dt = d(H_\alpha - f_{\alpha\beta})\, dt = (dH_\alpha - \dot{f}_{\alpha\beta}\, dt)\, dt = dH_\alpha\, dt.$$

We shall call this 2-form $d\omega$, even though there is no one-form ω on all of $\mathbf{S}$ which it is the "d" of.

Suppose we have on some portion of $\mathbf{S}$ an r-parameter family of solutions of the equations of motion. Let us denote by $x_1, \cdots, x_r$ the parameters. What this means is that we have a mapping ϕ on a region $\mathbf{W}$ of $(t, \mathbf{x})$ space of the sort indicated, a cylinder in the t direction with top and bottom curved r-chains,

$$\phi\colon\ \mathbf{W} \longrightarrow \mathbf{S}$$

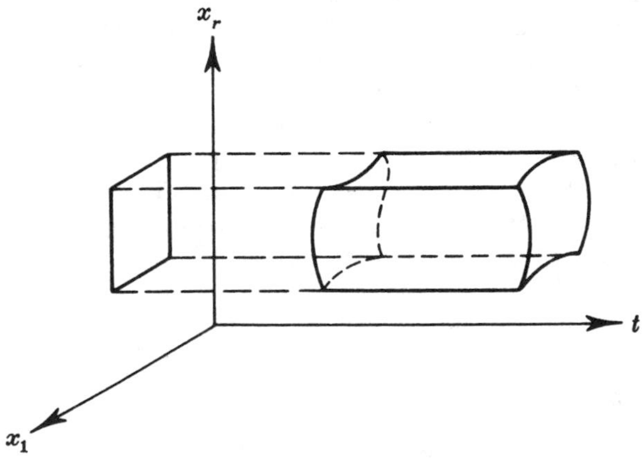

where in local coordinates

$$\phi: \begin{cases} q^i = f^i(t, x_1, \cdots, x_r) \\ p_i = g_i(t, x_1, \cdots, x_r) \\ t = t. \end{cases}$$

For each (x_i), this represents a trajectory, hence

$$\begin{cases} \dfrac{\partial f^i}{\partial t} = \dfrac{\partial H}{\partial p_i}(\mathbf{f}, \mathbf{g}, t) \\ \dfrac{\partial g_i}{\partial t} = -\dfrac{\partial H}{\partial q^i}(\mathbf{f}, \mathbf{g}, t). \end{cases}$$

The mapping ϕ is supposed smooth and one-to-one, so that an $(r+1)$-dimensional region in $\mathbf{S}$ is evenly filled up by these trajectories.

Now we compute $\phi^*(d\omega)$. We have

$$\begin{aligned} d\omega &= \sum dp_i dq^i - \sum \frac{\partial H}{\partial q^i} dq^i dt - \sum \frac{\partial H}{\partial p_i} dp_i dt \\ &= \sum \left(dp_i + \frac{\partial H}{\partial q^i} dt\right)\left(dq^i - \frac{\partial H}{\partial p_i} dt\right). \end{aligned}$$

Now

$$\phi^*\left(dq^i - \frac{\partial H}{\partial p_i} dt\right) = \left(\frac{\partial f^i}{\partial t} dt + \sum \frac{\partial f^i}{\partial x_j} dx_j\right) - \frac{\partial f^i}{\partial t} dt = \sum \frac{\partial f^i}{\partial x_j} dx_j$$

and similarly

$$\phi^*\left(dp_i + \frac{\partial H}{\partial q^i} dt\right) = \sum \frac{\partial q^i}{\partial x_k} dx_k,$$

so that

$$\begin{aligned} \phi^*(d\omega) &= \tfrac{1}{2} \sum \frac{\partial(g_i, f^i)}{\partial(x_j, x_k)} dx_j dx_k \\ &= \sum A^{jk}(\mathbf{x}, t)\, dx_j dx_k, \end{aligned}$$

which establishes our first point, *$\phi^*(d\omega)$ is independent of dt.*

Since $d(d\omega) = 0$, we have $d[\phi^*(d\omega)] = 0$. But

$$d[\phi^*(d\omega)] = \sum \frac{\partial A^{jk}}{\partial t} dt\, dx_j dx_k + (\text{terms in } dx_i dx_j dx_k) = 0.$$

We conclude that

$$\frac{\partial A^{jk}}{\partial t} = 0,$$

each $A^{jk} = A^{jk}(\mathbf{x})$ *is independent of* t, and we may write

$$\phi^*(d\omega) = \sum A^{jk}(\mathbf{x})\, dx_j\, dx_k .$$

A differential form α of degree s on the state space $\mathbf{S}$ is called an (*absolute*) *integral-invariant* (historical terminology) if for each r-parameter family of trajectories given by such a mapping ϕ, $\phi^*(\alpha)$ is an s-form on the $\mathbf{x}$-space alone (no t nor dt terms) and if in addition $d\alpha = 0$.

Each of the forms

$$d\omega,\ (d\omega)^2,\ \cdots,\ (d\omega)^n$$

is an integral-invariant. For

$$d(d\omega)^s = 0$$

and

$$\phi^*(d\omega)^s = [\phi^*(d\omega)]^s = [\sum A^{jk}(\mathbf{x})\, dx_j\, dx_k]^s$$

is independent of t and dt.

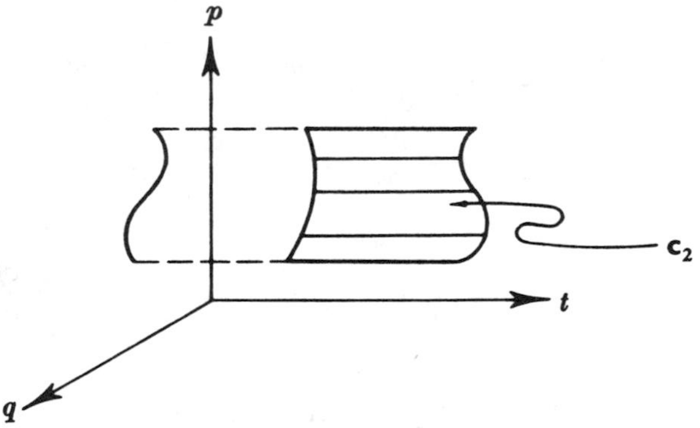

Consider in state space $\mathbf{S}$ a small piece $\mathbf{c}_2$ of surface which is filled by a one-parameter family of trajectories. We may describe $\mathbf{c}_2$ analytically by

$$\begin{cases} q^i = q^i(t, y) \\ p_i = p_i(t, y) \end{cases}$$

$$\begin{cases} a(y) \leqq t \leqq b(y) \\ \quad y_0 \leqq y \leqq y_1 . \end{cases}$$

Our reasoning above shows that the two-form we get by substituting these expressions for p and q in $d\omega$ is a two-form in y and dy only, hence vanishes. This means in particular

$$\int_{\mathbf{c}_2} d\omega = 0.$$

Next, suppose one has a piece of volume, or three-chain $\mathbf{c}_3$ in $\mathbf{S}$ which is the span of a two-parameter family of trajectories:

$$\begin{cases} q^i = q^i(t, y, z) \\ p_i = p_i(t, y, z) \end{cases}$$

$$\begin{cases} a(y, z) \leqq t \leqq b(y, z) \\ (y, z) \text{ in a domain } \mathbf{D}. \end{cases}$$

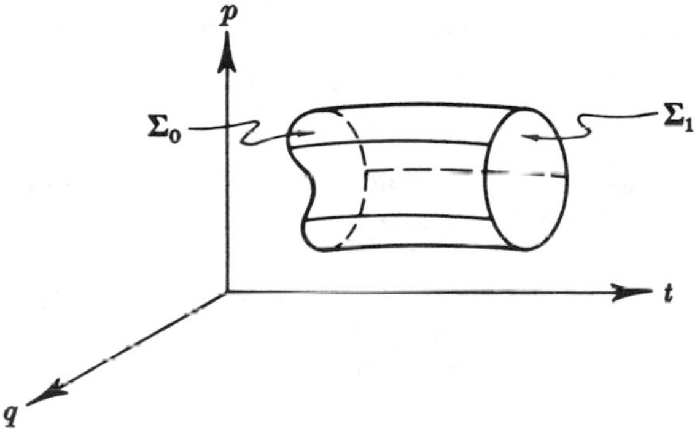

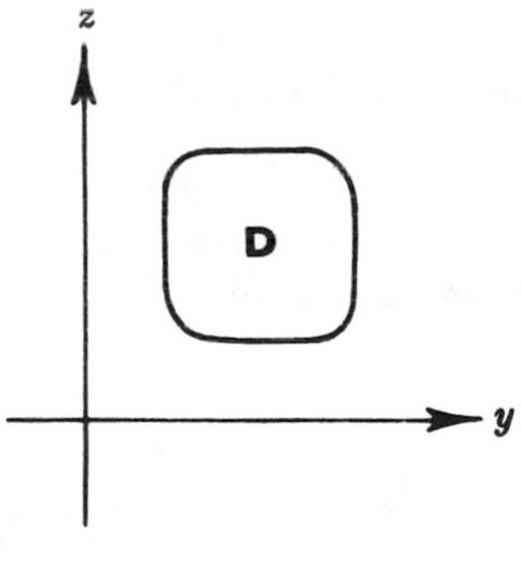

Then

$$\partial \mathbf{c}_3 = \Sigma_1 - \Sigma_0 + \mathbf{c}_2$$

where Σ_1, Σ_0 are the terminal and initial surfaces respectively and where the lateral surface $\mathbf{c}_2$ is spanned by a one-parameter family of trajectories corresponding to the parameter point (y, z) on $\partial\mathbf{D}$. Now

$$\int_{\partial \mathbf{c}_3} d\omega = \int_{\mathbf{c}_3} d(d\omega) = 0,$$

and we showed above that

$$\int_{\mathbf{c}_2} d\omega = 0,$$

hence

$$\int_{\Sigma_1} d\omega = \int_{\Sigma_0} d\omega.$$

We have established the very striking property of the form $d\omega$:

If Σ_0 is any 2-chain in **S** *transversal to the trajectories, and if one displaces each point of Σ_0 any amount along its trajectory to form a new surface Σ_1, then*

$$\int_{\Sigma_1} d\omega = \int_{\Sigma_0} d\omega.$$

It should be clear that we have not used any special properties of $d\omega$ in proving this result so that any integral-invariant satisfies a corresponding property.

Let α be an integral-invariant of degree r on **S**. *Let $\mathbf{c}_r$ be any r-chain on* **S** *transversal to the trajectories. Let $\mathbf{c}'_r$ be a second such r-chain so that the points of $\mathbf{c}_r$ and $\mathbf{c}'_r$ may be put into one-one correspondence with corresponding points on the same trajectory. Then*

$$\int_{\mathbf{c}_r} \alpha = \int_{\mathbf{c}'_r} \alpha.$$

It is possible to reverse our steps to prove that the property expressed in this result actually characterizes integral-invariants. This is done in É. Cartan [8].

We pass on to relative integral-invariants.

An r-form α on **S** is a *relative integral-invariant* provided $d\alpha$ is an integral-invariant. The basic result about relative integral-invariants is this.

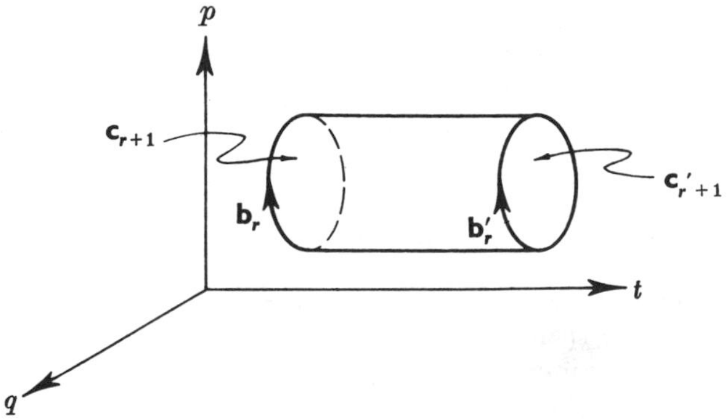

Let α be a relative integral-invariant of degree r. Let $\mathbf{b}_r$ and $\mathbf{b}'_r$ be two r-dimensional boundaries which are in one-one correspondence in such a way that corresponding points are on the same trajectory. Then

$$\int_{\mathbf{b}_r} \alpha = \int_{\mathbf{b}'_r} \alpha.$$

To prove this, we select $(r + 1)$-chains $\mathbf{c}_{r+1}$, $\mathbf{c}'_{r+1}$ so that

$$\mathbf{b}_r = \partial \mathbf{c}_{r+1}, \qquad \mathbf{b}'_r = \partial \mathbf{c}'_{r+1}$$

and do this in such a way that $\mathbf{c}_{r+1}$ and $\mathbf{c}'_{r+1}$ correspond one-one with corresponding points on the same trajectory. Then

$$\int_{\mathbf{b}'_r} \alpha = \int_{\partial \mathbf{c}'_{r+1}} \alpha = \int_{\mathbf{c}'_{r+1}} d\alpha = \int_{\mathbf{c}_{r+1}} d\alpha = \int_{\partial \mathbf{c}_{r+1}} \alpha = \int_{\mathbf{b}_r} \alpha.$$

The differential form

$$\omega = \sum p_i \, dq^i - H \, dt,$$

which is defined locally, is a relative integral-invariant of degree one where it is defined. Consequently so are the forms

$$\omega \, d\omega, \; \omega(d\omega)^2, \cdots, \omega(d\omega)^n$$

since

$$d[\omega(d\omega)^r] = (d\omega)^{r+1}.$$

We shall specialize by considering chains which exist at a single instance of time. Let α be any differential r-form and $\mathbf{c}$ an r-chain in $\mathbf{S}$ lying in a hyperplane $t =$ constant. Then clearly

$$\int_{\mathbf{c}} \alpha = \int_{\mathbf{c}} \alpha \,|_{dt=0} \,.$$

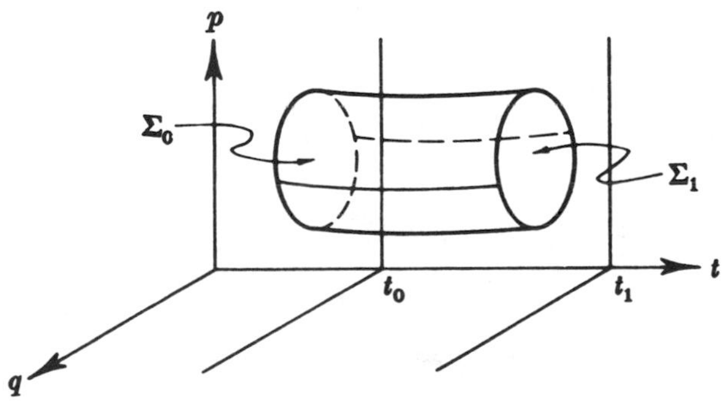

Let us apply this to $d\omega$ in particular. We have

$$d\omega\,|_{dt=0} = \sum dp_i\,dq^i,$$

which leads to this result:

Let Σ_0 be a 2-chain in $\mathbf{S}$ *at $t = t_0$ and Σ_1 the 2-chain obtained by moving each point of Σ_0 along its trajectory to time $t = t_1$. Then*

$$\int_{\Sigma_0} \sum dp_i\,dq^i = \int_{\Sigma_1} \sum dp_i\,dq^i.$$

In this result we may think of Σ_0 and Σ_1 as 2-chains in phase space.

If we apply the procedure to the $(2n)$-form

$$(d\omega)^n = \pm\, n!(dp_1 \cdots dp_n\,dq^1 \cdots dq^n) + \lambda\,dt$$

which gives us the *phase density*

$$\mu = dp_1 \cdots dp_n\,dq^1 \cdots dq^n$$

we have the

LIOUVILLE THEOREM. *If a 2n-dimensional region* $\mathbf{D}_0$ *in phase space at time t_0 moves to a region* $\mathbf{D}_1$ *at time t_1*, then

$$\int_{\mathbf{D}_1} \mu = \int_{\mathbf{D}_0} \mu.$$

We shall close this section with a result which will be needed in Section 10.5. It shows that the integral-invariant $d\omega$ completely determines the equations of motion, in itself an important mechanical principle.

Let

$$\begin{cases} \dot{q}^i = A^i(t, \mathbf{q}, \mathbf{p}) \\ \dot{p}_i = B_i(t, \mathbf{q}, \mathbf{p}) \end{cases}$$

be a system of equations on a region of state space $\mathbf{S}$ *which has*

$$d\omega = \sum dp_i\,dq^i - dH dt$$

as an integral-invariant. Then

$$A^i = \frac{\partial H}{\partial p_i}, \qquad B_i = -\frac{\partial H}{\partial q^i}.$$

For let

$$\begin{cases} q^i = f^i(t, x_1, \cdots, x_{2n}) \\ p_i = g_i(t, x_1, \cdots, x_{2n}) \end{cases}$$

be a general solution, so that $d\omega$ must be expressible in the terms of dx_i alone. From this, and the differential equations

$$\left\{\frac{\partial f^i}{\partial t} = A^i, \qquad \frac{\partial g_i}{\partial t} = B_i\right\}$$

we deduce

$$dq^i = A^i\,dt + \lambda^i, \qquad dp_i = B_i\,dt + \mu_i,$$

where λ^i, μ_i are one-forms in the dx_j alone, and so

$$\begin{aligned} d\omega &= \sum dp_i\,dq^i - dH\,dt \\ &= \left(\sum A^i\mu_i - \sum B_i\lambda^i\right)dt + \sum \mu_i\lambda^i - \left(\sum \frac{\partial H}{\partial q^i}\lambda^i + \sum \frac{\partial H}{\partial p_i}\mu_i\right)dt. \end{aligned}$$

Since $d\omega$ is free of dt,

$$\sum A^i\mu_i - \sum B_i\lambda^i - \sum \frac{\partial H}{\partial q^i}\lambda^i - \sum \frac{\partial H}{\partial p_i}\mu_i = 0,$$

$$\sum \left(A^i - \frac{\partial H}{\partial p_i}\right)\mu_i - \sum \left(B_i + \frac{\partial H}{\partial q^i}\right)\lambda^i = 0,$$

$$\sum \left(A^i - \frac{\partial H}{\partial p_i}\right)(dp_i - B_i\,dt) - \sum \left(B_i + \frac{\partial H}{\partial q^i}\right)(dq^i - A^i\,dt) = 0,$$

$$\sum \left(A^i - \frac{\partial H}{\partial p_i}\right)dp_i - \sum \left(B_i + \frac{\partial H}{\partial q^i}\right)dq^i + (\cdots)\,dt = 0.$$

We conclude that

$$A^i - \frac{\partial H}{\partial p_i} = 0, \qquad B_i + \frac{\partial H}{\partial q^i} = 0$$

as asserted.

10.4. Brackets

In the transformation theory of classical mechanics one uses the bracket expressions of Poisson and Lagrange. In this section we shall show how these expressions relate to differential forms.

Before doing this, it is a good idea to digress on the subject of Lie brackets. We take any differentiable manifold $\mathbf{N}$ and recall the definition in Section 5.3 of tangent vector and the definition of vector field at the end of that section. We may consider a vector field $\mathbf{v}$ on $\mathbf{N}$ as an operator which takes each function on $\mathbf{N}$ to another function on $\mathbf{N}$:

$$\mathbf{v}\colon \mathbf{F}^0(\mathbf{N}) \longrightarrow \mathbf{F}^0(\mathbf{N}).$$

If $x^1, \cdots, x^n$ is a local coordinate system in which $\mathbf{v}$ has the representation

$$\mathbf{v} = \sum a^i(\mathbf{x})\,\frac{\partial}{\partial x^i},$$

then

$$\mathbf{v}(f) = \sum a^i(\mathbf{x})\,\frac{\partial f}{\partial x^i}$$

shows, locally, what $\mathbf{v}$ does to a function f. One sees from this formula, or from the very definition of $\mathbf{v}$ as the assignment of a tangent vector (directional differentiation) at P to each point P of $\mathbf{N}$, that

$$\mathbf{v}(f \cdot g) = \mathbf{v}(f) \cdot g + f \cdot \mathbf{v}(g)$$

for any two functions f and g on $\mathbf{N}$. If $\mathbf{v}$ and $\mathbf{w}$ are two vector fields on $\mathbf{N}$, we define the *Lie bracket* of $\mathbf{v}$ and $\mathbf{w}$ by

$$[\mathbf{v}, \mathbf{w}] = \mathbf{v} \circ \mathbf{w} - \mathbf{w} \circ \mathbf{v}.$$

This is another vector field on $\mathbf{N}$. If in local coordinates

$$\mathbf{v} = \sum a^i \frac{\partial}{\partial x^i}, \qquad \mathbf{w} = \sum b^j \frac{\partial}{\partial x^j},$$

then a computation shows that

$$[\mathbf{v}, \mathbf{w}] = \sum \left(\sum a^i \frac{\partial b^j}{\partial x^i} - \sum b^i \frac{\partial a^j}{\partial x^i} \right) \frac{\partial}{\partial x^j}.$$

(The main point to notice is that the second partials cancel each other.) The following algebraic identities are easily established:

(i) $[\mathbf{v}, \mathbf{v}] = 0.$

(ii) $[\mathbf{w}, \mathbf{v}] + [\mathbf{v}, \mathbf{w}] = 0.$

(iii) $[\mathbf{v}_1 + \mathbf{v}_2, \mathbf{w}] = [\mathbf{v}_1, \mathbf{w}] + [\mathbf{v}_2, \mathbf{w}].$

(iv) $\big[[\mathbf{u}, \mathbf{v}], \mathbf{w}\big] + \big[[\mathbf{v}, \mathbf{w}], \mathbf{u}\big] + \big[[\mathbf{w}, \mathbf{u}], \mathbf{v}\big] = 0$ (Jacobi identity).

Now we return to mechanical systems. As before, let $\mathbf{P}$ be the phase space associated to a position space $\mathbf{M}$. We denote by α the differential form

$$\alpha = \sum p_i \, dq^i$$

on $\mathbf{P}$ so that

$$d\alpha = \sum dp_i \, dq^i$$

as in Section 10.1.

Poisson Brackets. To each pair f, g of real functions on phase space P we associate a new function (f, g) defined by

$$n(df\,dg) \wedge (d\alpha)^{n-1} = (f, g)(d\alpha)^n.$$

In local coordinates

$$(d\alpha)^{n-1} = [(n-1)!] \sum (dp_1\, dq^1) \cdots \widehat{(dp_i\, dq^i)} \cdots (dp_n\, dq^n)$$

and

$$(d\alpha)^n = (n!)\, dp_1\, dq^1 \cdots dp_n\, dq^n,$$

from which we deduce the local expression for (f, g):

$$(f, g) = \sum \frac{\partial(f, g)}{\partial(p_i, q^i)}.$$

From the definition one derives these relations:

(i) $(f, f) = 0$.

(ii) $(f, g) + (g, f) = 0$.

(iii) $(f, g_1 + g_2) = (f, g_1) + (f, g_2)$.

Using $(df)\, d(g_1 g_2) = g_1(df\, dg_2) + g_2(df\, dg_1)$, one has

(iv) $(f, g_1 g_2) = g_1 \cdot (f, g_2) + g_2 \cdot (f, g_1)$.

The identities (iii) and (iv) taken together may be expressed by saying that for fixed f,

$$g \longrightarrow (f, g)$$

is a vector field $\mathbf{v}_f$ on P:

$$\mathbf{v}_f(g) = (f, g).$$

The basic connection between the Lie and Poisson brackets is given by

(v) $\mathbf{v}_{(f, g)} = [\mathbf{v}_f, \mathbf{v}_g]$.

One has

$$\mathbf{v}_{(f, g)}(h) = \big((f, g), h\big)$$

and

$$\begin{aligned} [\mathbf{v}_f, \mathbf{v}_g](h) &= \mathbf{v}_f[\mathbf{v}_g(h)] - \mathbf{v}_g[\mathbf{v}_f(h)] \\ &= \mathbf{v}_f\big((g, h)\big) - \mathbf{v}_g\big((f, h)\big) \\ &= \big(f, (g, h)\big) - \big(g, (f, h)\big) \\ &= -\big((g, h), f\big) - \big((h, f), g\big) \end{aligned}$$

so that (v) is equivalent to *Jacobi's relation*

(vi) $\big((f, g), h\big) + \big((g, h), f\big) + \big((h, f), g\big) = 0$.

One proves this relation [(v) or (vi)] either by a lengthy direct calculation, or by the following more sophisticated argument based on the fact that a vector field is completely determined locally by its effect on each member of a local coordinate system.

First of all,

$$(f, q^i) = \sum \frac{\partial(f, q^i)}{\partial(p_j, q^j)} = \sum \begin{vmatrix} \dfrac{\partial f}{\partial p_j} & \dfrac{\partial f}{\partial q^j} \\ 0 & \delta^i_j \end{vmatrix},$$

$$(f, q^i) = \frac{\partial f}{\partial p_i},$$

which may be interpreted as

$$\mathbf{v}_{q^i} = -\frac{\partial}{\partial p^i}.$$

Similarly

$$(f, p_i) = -\frac{\partial f}{\partial q^i} \qquad \text{and} \qquad \mathbf{v}_{p_i} = \frac{\partial}{\partial q^i}.$$

Because of these relations (vi) easily follows when *both* g and h are taken from the set of coordinate functions $\{q^1, \cdots, p_n\}$. This means that

$$\mathbf{v}_{(f,x)} = [\mathbf{v}_f, \mathbf{v}_x]$$

when x is one of these coordinate functions since the vector fields on both sides agree when applied to any coordinate function q^j or p_j. Hence for any h,

$$\big((f, x), h\big) + \big((x, h), f\big) + \big((h, f), x\big) = 0.$$

This is now established for any functions f and h and any coordinate function x. But this may be interpreted as saying

$$\mathbf{v}_{(h,f)}x = [\mathbf{v}_h, \mathbf{v}_f]x$$

so that the vector fields

$$\mathbf{v}_{(h,f)} \qquad \text{and} \qquad [\mathbf{v}_h, \mathbf{v}_f]$$

must agree since they agree on all of the coordinate functions. Hence

$$\mathbf{v}_{(h,f)}(g) = [\mathbf{v}_h, \mathbf{v}_f]g$$

for all f, g, h. This completes the proof of (v) and (vi).

One applies brackets to a pair of functions on the state space **S** by simply treating t as a parameter.

A function f on **S** is called a *first integral* of the equations of motion if it is constant on each trajectory. If this is the case, then

$$\begin{aligned} 0 = \frac{df}{dt} &= \sum \frac{\partial f}{\partial q^i}\dot{q}^i + \sum \frac{\partial f}{\partial p_i}\dot{p}_i + \frac{\partial f}{\partial t} \\ &= \sum \frac{\partial f}{\partial q^i}\frac{\partial H}{\partial p_i} - \sum \frac{\partial f}{\partial p_i}\frac{\partial H}{\partial q^i} + \frac{\partial f}{\partial t} \\ &= (H, f) + \frac{\partial f}{\partial t} \end{aligned}$$

so that the partial differential equation for first integrals is

$$\frac{\partial f}{\partial t} = (f, H).$$

LAGRANGE BRACKETS. Let

$$\phi\colon \quad \mathbf{E}^2 \longrightarrow \mathbf{P},$$

where $\mathbf{E}^2$ is the Euclidean plane with rectangular coordinates u, v. Then $\phi^*(d\alpha)$ is a 2-form on $\mathbf{E}^2$ (where as before $d\alpha = \sum dp_i dq^i$) and we write

$$\phi^*(d\alpha) = [u, v]\, du\, dv$$

defining the *Lagrange brackets* which have the local expression

$$[u, v] = \sum \frac{\partial(p_i, q^i)}{\partial(u, v)}.$$

10.5. Contact Transformations

Here we can but touch briefly on an extensive topic. For simplicity we shall only treat the local problem and shall look on contact transformations as coordinate changes.

First we take the case of a time independent change. As usual, let q^i, p_i denote local coordinates in phase space $\mathbf{P}$. We consider new coordinates $\bar{q}^i$, $\bar{p}_i$,

$$\begin{cases} \bar{q}^i = \bar{q}^i(q^1, \cdots, q^n, p_1, \cdots, p_n) \\ \bar{p}_i = \bar{p}_i(q^1, \cdots, q^n, p_1, \cdots, p_n) \end{cases}$$

which are unrelated to the old except for one requirement. We set

$$\alpha = \sum p_i dq^i, \qquad \bar{\alpha} = \sum \bar{p}_i d\bar{q}^i$$

and require

$$d\bar{\alpha} = d\alpha.$$

This defines what we shall call a *homogeneous contact transformation.*

Since $d(\bar{\alpha} - \alpha) = 0$ and we are only working locally, the condition may be expressed as

$$\bar{\alpha} = \alpha + d\phi$$

where ϕ is a real function on $\mathbf{P}$. In the relation $d\bar{\alpha} = d\alpha$,

$$\sum d\bar{p}_i d\bar{q}^i = \sum dp_i dq^i,$$

one replaces $d\bar{p}_i$ and $d\bar{q}^i$ by their expressions in terms of dp_j and dq^k to obtain

$$\left\{\sum \frac{\partial(\bar{p}_i, \bar{q}^i)}{\partial(q^j, q^k)} = 0, \qquad \sum \frac{\partial(\bar{p}_i, \bar{q}^i)}{\partial(p_j, p_k)} = 0, \qquad \sum \frac{\partial(\bar{p}_i, \bar{q}^i)}{\partial(p_j, q^k)} = \delta^j_k\right\}.$$

(These relations may be expressed in terms of Lagrange brackets.) Similarly,

$$\left\{\sum \bar{p}_i \frac{\partial \bar{q}^i}{\partial q^j} = p_j + \frac{\partial \phi}{\partial q^j}, \qquad \sum \bar{p}_i \frac{\partial \bar{q}^i}{\partial p_j} = \frac{\partial \phi}{\partial p_j}\right\}.$$

If we have equations of motion

$$\begin{cases} \dot{q}^i = \dfrac{\partial H}{\partial p_i} \\ \dot{p}_i = -\dfrac{\partial H}{\partial q^i} \end{cases}$$

we shall show that they transform into equations of the same type. For suppose after changing the coordinates according to our contact transformation we arrive at

$$\begin{cases} \dot{\bar{q}}^i = A^i \\ \dot{\bar{p}}_i = -B_i \end{cases}$$

so that

$$\begin{cases} A^i = \sum \dfrac{\partial \bar{q}^i}{\partial q^j}\dfrac{\partial H}{\partial p_j} - \sum \dfrac{\partial \bar{q}^i}{\partial p_j}\dfrac{\partial H}{\partial q^j} \\ B_i = -\sum \dfrac{\partial \bar{p}_i}{\partial q^j}\dfrac{\partial H}{\partial p_j} + \sum \dfrac{\partial \bar{p}_i}{\partial p_j}\dfrac{\partial H}{\partial q^j}\,. \end{cases}$$

If we multiply the first by

$$d\bar{p}_i = \sum \frac{\partial \bar{p}_i}{\partial q^k}\, dq^k + \sum \frac{\partial \bar{p}_i}{\partial p_k}\, dp_k$$

and the second by $d\bar{q}^i$ similarly and sum, we find after some simplification

$$\begin{aligned} \sum A^i d\bar{p}_i + \sum B_i d\bar{q}^i &= \sum \frac{\partial H}{\partial q^j}\, dq^j + \sum \frac{\partial H}{\partial p_k}\, dp_k \\ &= dH - \frac{\partial H}{\partial t}\, dt. \end{aligned}$$

Hence

$$A^i = \frac{\partial H}{\partial \bar{p}_i}, \qquad B_i = \frac{\partial H}{\partial \bar{q}^i}$$

so that the equations of motion in the new coordinates are precisely

$$\begin{cases} \dot{\bar{q}}^i = \dfrac{\partial H}{\partial \bar{p}_i} \\ \dot{\bar{p}}_i = -\dfrac{\partial H}{\partial \bar{q}^i}\,. \end{cases}$$

Now we pass to the general situation. We begin with a mechanical system

$$\begin{cases} \dot{q}^i = \dfrac{\partial H}{\partial p_i} \\ \dot{p}_i = -\dfrac{\partial H}{\partial q^i} \end{cases}$$

on the state space **S** and consider a coordinate change on **S**:

$$\begin{cases} \bar{q}^i = \bar{q}^i(t, q^1, \cdots, q^n, p_1, \cdots, p_n) \\ \bar{p}_i = \bar{p}_i(t, q^1, \cdots, q^n, p_1, \cdots, p_n) \\ t = t. \end{cases}$$

We set

$$\omega = \sum p_i \, dq^i - H \, dt$$

as usual. The coordinate change is called a *contact transformation* if there is a function $\overline{H}$ and a function ϕ so that if

$$\bar{\omega} = \sum \bar{p}_i \, d\bar{q}^i - \overline{H} \, dt,$$

then

$$\bar{\omega} = \omega + d\phi,$$

or what is the same thing,

$$d\bar{\omega} = d\omega.$$

The first basic result is that the equations of motion in the new coordinates are precisely

$$\begin{cases} \dot{\bar{q}}^i = \dfrac{\partial \overline{H}}{\partial \bar{p}_i} \\ \dot{\bar{p}}_i = -\dfrac{\partial \overline{H}}{\partial \bar{q}^i}. \end{cases}$$

For whatever the new equations of motion are, they admit

$$d\bar{\omega} = \sum d\bar{p}_i \, d\bar{q}^i - d\overline{H} \, dt$$

as an integral-invariant. But the final result in Section 10.3 tells us that the only equations with this integral-invariant are the stated ones. (Compare this slick proof with a direct computation!)

A particular type of contact transformation is obtained as follows. Let

$$\phi(t, x^1, \cdots, x^n, y^1, \cdots, y^n)$$

be a function of $2n + 1$ variables satisfying the independence condition

$$\left| \frac{\partial^2 \phi}{\partial x^i \, \partial y^j} \right| \neq 0.$$

Set $\phi = \phi(t, q^1, \cdots, q^n, \bar{q}^1, \cdots, \bar{q}^n)$,

$$p_i = \frac{\partial\phi}{\partial q^i}, \qquad \bar{p}_i = -\frac{\partial\phi}{\partial \bar{q}^i}.$$

For fixed q and p, the $\bar{q}$ are determined by the first set of equations (since the determinant above does not vanish) and then the $\bar{p}$ are determined by the second set of equations.

We have

$$d\phi = \frac{\partial\phi}{\partial t}\,dt + \sum \frac{\partial\phi}{\partial q^i}\,dq^i + \sum \frac{\partial\phi}{\partial \bar{q}^i}\,d\bar{q}^i$$

$$= \frac{\partial\phi}{\partial t}\,dt + \sum p_i\,dq^i - \sum \bar{p}_i\,d\bar{q}^i,$$

$$\omega - \bar{\omega} = -\left(\sum \bar{p}_i\,d\bar{q}^i - \bar{H}\,dt\right) + \left(\sum p_i\,dq^i - H\,dt\right)$$

$$= +d\phi - \frac{\partial\phi}{\partial t}\,dt + \bar{H}\,dt - H\,dt,$$

and so

$$\bar{\omega} = \omega - d\phi$$

provided we set

$$\bar{H} = H + \frac{\partial\phi}{\partial t}.$$

The most important case is that in which ϕ is a solution of the *Hamilton-Jacobi equation*

$$\frac{\partial\phi}{\partial t}(t, q, \bar{q}) + H\left(t, q, \cdots, \frac{\partial\phi}{\partial q^i}, \cdots\right) = 0.$$

In this situation $\bar{H} = 0$ and the new equations of motion are simply

$$\begin{cases} \dot{\bar{q}}^i = 0 \\ \dot{\bar{p}}_i = 0 \end{cases}$$

with solutions

$$\bar{q}^i = \text{constant}, \qquad \bar{p}_i = \text{constant}.$$

The original system is said to be *transformed to equilibrium.*

One other point we shall notice is this. If a contact transformation is stationary, i.e., independent of time, then it is equivalent to a homogeneous contact transformation. For suppose

$$\begin{cases} \bar{q}^i = \bar{q}^i(q, p) \\ \bar{p}_i = \bar{p}_i(q, p) \end{cases}$$

and

$$\left(\sum \bar{p}_i\,d\bar{q}^i - \bar{H}\,dt\right) = \left(\sum p_i\,dq^i - H\,dt\right) + d\phi.$$

Equating coefficients:

$$\begin{cases} \sum \bar{p}_i \dfrac{\partial \bar{q}^i}{\partial q^j} = p_j + \dfrac{\partial \phi}{\partial q^j} \\ \sum \bar{p}_i \dfrac{\partial \bar{q}^i}{\partial p_k} = \dfrac{\partial \phi}{\partial p_k} \\ \qquad \overline{H} = H - \dfrac{\partial \phi}{\partial t}. \end{cases}$$

Since q, p, $\bar{q}$, $\bar{p}$ are independent of t, we deduce from the first two equations that

$$\frac{\partial^2 \phi}{\partial t\, \partial q^j} = 0, \qquad \frac{\partial^2 \phi}{\partial t\, \partial p_k} = 0.$$

Hence $\partial \phi / \partial t$ is a function of t alone. This evidently implies

$$\phi(t, q, p) = f(t) + \psi(q, p)$$

and so

$$\sum \bar{p}_i d\bar{q}^i = \sum p_i dq^i + d\psi$$

which means we have a homogeneous contact transformation as asserted.

Let us briefly examine what are called *infinitesimal contact transformations*. We ask when

$$\begin{cases} \bar{q}^i = q^i + \varepsilon f^i \\ \bar{p}_i = p_i + \varepsilon g_i \end{cases}$$

is a contact transformation up to first order terms in ε. We restrict attention to the homogeneous (stationary) case. We have

$$\sum \bar{p}_i d\bar{q}^i - \sum p_i dq^i = \varepsilon\, d\phi,$$

$$\sum (p_i + \varepsilon g_i)(dq^i + \varepsilon\, df^i) - \sum p_i dq^i = \varepsilon\, d\phi,$$

$$\varepsilon\left(\sum g_i dq^i + \sum p_i df^i\right) = \varepsilon\, d\phi,$$

so the condition is

$$\sum g_i dq^i + \sum p_i df^i = d\phi.$$

If we set

$$\psi = \phi - \sum p_i f^i$$

this becomes

$$\sum g_i dq^i - \sum f^i dp_i = d\psi,$$

or

$$g_i = \frac{\partial \psi}{\partial q^i}, \qquad f^i = -\frac{\partial \psi}{\partial p_i}.$$

We finally have

$$\bar{q}^i = q^i - \varepsilon \frac{\partial \psi}{\partial p_i}$$

$$\bar{p}_i = p_i + \varepsilon \frac{\partial \psi}{\partial q^i} .$$

Here ψ is called a *generating function* for the infinitesimal contact transformation.

10.6. Fluid Mechanics

We consider a fluid moving in a region of $\mathbf{E}^3$. The position vector as usual is

$$\mathbf{x} = (x, y, z) = (x^1, x^2, x^3).$$

At each time t, the velocity at $\mathbf{x}$ is $\mathbf{v}$,

$$\mathbf{v} = \mathbf{v}(t, \mathbf{x}) = (u, v, w) = (v^1, v^2, v^3).$$

The density of the fluid is a scalar

$$\rho = \rho(t, \mathbf{x}).$$

In this section we shall denote the vectorial area element of a surface by $\boldsymbol{\sigma}$,

$$\boldsymbol{\sigma} = (dy\,dz,\ dz\,dx,\ dx\,dy).$$

(See Section 4.5, p. 43.)

If $\mathbf{c}_3$ is a three-dimensional region which is fixed in space, the change in mass at each point $\mathbf{x}$ of $\mathbf{c}_3$ per unit time is

$$\frac{\partial \rho}{\partial t}\, dx\, dy\, dz$$

and so the total time derivative of mass in $\mathbf{c}_3$ is

$$\int_{\mathbf{c}_3} \frac{\partial \rho}{\partial t}\, dx\, dy\, dz.$$

We assume conservation of matter, so this must result from flux of fluid over the boundary, hence

$$\int_{\mathbf{c}_3} \frac{\partial \rho}{\partial t}\, dx\, dy\, dz = - \int_{\partial \mathbf{c}_3} \rho \mathbf{v} \cdot \boldsymbol{\sigma}.$$

By Gauss' theorem,

$$\int_{\partial \mathbf{c}_3} \rho \mathbf{v} \cdot \boldsymbol{\sigma} = \int_{\mathbf{c}_3} \operatorname{div}(\rho \mathbf{v})\, dx\, dy\, dz.$$

By taking $\mathbf{c}_3$ arbitrary we deduce from the equality of these integrals the *continuity equation*

$$\frac{\partial \rho}{\partial t} + \operatorname{div}(\rho \mathbf{v}) = 0,$$

a necessary condition the flow must satisfy. We shall deduce some consequences of this. We set

$$\Omega = \rho(dx^1 - v^1\,dt)(dx^2 - v^2\,dt)(dx^3 - v^3\,dt).$$

To compute $d\Omega$, we set

$$\beta = (dx^1 - v^1\,dt)(dx^2 - v^2\,dt)(dx^3 - v^3\,dt)$$

so that

$$\begin{aligned} d\beta &= -\frac{\partial v^1}{\partial x^1}\,dx^1\,dt\,dx^2\,dx^3 + dx^1\left(\frac{\partial v^2}{\partial x^2}\,dx^2\,dt\right)dx^3 - dx^1\,dx^2\left(\frac{\partial v^3}{\partial x^3}\,dx^3\,dt\right) \\ &= (\operatorname{div}\mathbf{v})\,dt\,dx^1\,dx^2\,dx^3, \\ d\Omega &= d(\rho\beta) = d\rho \wedge \beta + \rho\,d\beta \\ &= \left(\frac{\partial \rho}{\partial t}\,dt + \sum \frac{\partial \rho}{\partial x^i}\,dx^i\right) \wedge \beta + \rho(\operatorname{div}\mathbf{v})(dt\,dx^1\,dx^2\,dx^3) \\ &= \left[\frac{\partial \rho}{\partial t} + \sum v^i\frac{\partial \rho}{\partial x^i} + \rho(\operatorname{div}\mathbf{v})\right](dt\,dx^1\,dx^2\,dx^3). \end{aligned}$$

Thus the continuity equation is equivalent to the relation

$$d\Omega = 0.$$

Suppose we express the flow in terms of initial conditions (or other parameters) by

$$\mathbf{x} = \mathbf{x}(t, \alpha^1, \cdots, \alpha^3)$$

so that the α^i are the parameters and

$$\frac{\partial \mathbf{x}}{\partial t} = \mathbf{v}.$$

Thus

$$\begin{aligned} (dx^i - v^i\,dt) &= \left(\frac{\partial x^i}{\partial t}\,dt + \sum \frac{\partial x^i}{\partial \alpha^j}\,d\alpha^j\right) - v^i\,dt \\ &= \sum \frac{\partial x^i}{\partial \alpha^j}\,d\alpha^j \end{aligned}$$

so that

$$\begin{aligned} \Omega &= \rho\,\frac{\partial(x^1, x^2, x^3)}{\partial(\alpha^1, \alpha^2, \alpha^3)}\,d\alpha^1\,d\alpha^2\,d\alpha^3 \\ &= A(t, \boldsymbol{\alpha})d\alpha^1\,d\alpha^2\,d\alpha^3. \end{aligned}$$

Since $d\Omega = 0$, we deduce that $\partial A/\partial t = 0$,

$$\Omega = A(\boldsymbol{\alpha})\, d\alpha^1\, d\alpha^2\, d\alpha^3.$$

This means that Ω is an integral-invariant for the flow as explained in Section 10.3. We consequently have the following result.

Let $\mathbf{c}_3$, $\mathbf{c}'_3$ *be three-chains in the four-dimensional* $(t, \mathbf{x})$ *space which are in one-one correspondence in such a way that corresponding points lie on the same trajectory of the flow. Then*

$$\int_{\mathbf{c}_3} \Omega = \int_{\mathbf{c}'_3} \Omega.$$

In particular, if all the points of $\mathbf{c}_3$ exist simultaneously, i.e., at a fixed time t_0, then

$$\int_{\mathbf{c}_3} \Omega = \int_{\mathbf{c}_3} \Omega|_{t=t_0} = \int_{\mathbf{c}_3} \rho\, dx\, dy\, dz.$$

Thus if we take a region $\mathbf{c}_3^{(0)}$ at time t_0 and follow it to $\mathbf{c}_3^{(1)}$ at time t_1, we have

$$\int_{\mathbf{c}_3^{(0)}} \rho\, dx\, dy\, dz = \int_{\mathbf{c}_3^{(1)}} \rho\, dx\, dy\, dz$$

which says that mass is preserved in the flow, another form of the conservation of mass.

We now proceed to the dynamic situation. We suppose our fluid is nonviscous so that the pressure is a force per unit area at each point acting normal to any surface element through the point, always with the same magnitude. Let

$$p = p(t, \mathbf{x}) = \text{pressure}$$

$$\mathbf{F} = \mathbf{F}(t, \mathbf{x}) = \text{body force per unit mass.}$$

Let $\mathbf{c}_3$ be a fixed region in space. The total acceleration of all matter in $\mathbf{c}_3$ is

$$\int_{\mathbf{c}_3} \rho \frac{d\mathbf{v}}{dt}\, dx\, dy\, dz.$$

At the instant of time in question, this must equal the total force on the matter in $\mathbf{c}_3$ which is

$$\int_{\mathbf{c}_3} \rho \mathbf{F}\, dx\, dy\, dz - \int_{\partial \mathbf{c}_3} p\sigma.$$

By a variant† of Gauss' theorem,

$$\int_{\partial \mathbf{c}_3} p\sigma = \int_{\mathbf{c}_3} (\operatorname{grad} p)\, dx\, dy\, dz,$$

† The usual simple proof is to dot both sides into a constant vector $\mathbf{a}$ and apply Gauss' theorem plus the fact that

$$\operatorname{div}(p\mathbf{a}) = (\operatorname{grad} p)\cdot \mathbf{a}.$$

hence

$$\int_{\mathbf{c}_3} \left(\rho \frac{d\mathbf{v}}{dt} - \rho \mathbf{F} + \operatorname{grad} p\right) dx\, dy\, dz = 0.$$

We conclude that

$$\frac{d\mathbf{v}}{dt} = \mathbf{F} - \frac{1}{\rho} \operatorname{grad} p,$$

the *Euler equation of motion*.

Here the interpretation is

$$\frac{d\mathbf{v}}{dt} = \frac{\partial \mathbf{v}}{\partial t} + \sum \frac{\partial \mathbf{v}}{\partial x^i} v^i.$$

Let us suppose that the body force $\mathbf{F}$ is conservative,

$$\mathbf{F} = -\operatorname{grad} V$$

where

$$V = V(t, \mathbf{x})$$

is the force potential.

We shall add the hypothesis that p and ρ are functionally related, i.e.,

$$d\rho \wedge dp = 0,$$

as is the case, for example, with an isothermal motion. In this case we can define a function $q = q(t, \mathbf{x})$ by

$$q = \int_0^{(t, \mathbf{x})} \frac{dp}{\rho}$$

so that

$$dq = \frac{dp}{\rho}.$$

The equations of motion may be written

$$\frac{d\mathbf{v}}{dt} = -\operatorname{grad}(V + q)$$

or

$$\frac{du}{dt} = -\frac{\partial}{\partial x}(V + q), \qquad \text{etc.,}$$

i.e.,

$$\frac{\partial u}{\partial t} + u \frac{\partial u}{\partial x} + v \frac{\partial u}{\partial y} + w \frac{\partial u}{\partial z} = -\frac{\partial}{\partial x}(V + q), \qquad \text{etc.}$$

We set

$$\begin{aligned} E &= \tfrac{1}{2}(\mathbf{v} \cdot \mathbf{v}) + V + q \\ &= \tfrac{1}{2}(u^2 + v^2 + w^2) + V + q, \end{aligned}$$

the *energy per unit mass*. Finally we define the *vorticity*

$$\boldsymbol{\xi} = (\xi, \eta, \zeta) = \left(\frac{\partial w}{\partial y} - \frac{\partial v}{\partial z}, \quad \frac{\partial u}{\partial z} - \frac{\partial w}{\partial x}, \quad \frac{\partial v}{\partial x} - \frac{\partial u}{\partial y}\right).$$

We compute

$$\begin{aligned}\frac{\partial E}{\partial x} + \frac{\partial u}{\partial t} &= u\frac{\partial u}{\partial x} + \left(v\frac{\partial v}{\partial x} + w\frac{\partial w}{\partial x}\right) + \frac{\partial u}{\partial t} + \frac{\partial}{\partial x}(V + q)\\ &= \left(v\frac{\partial v}{\partial x} + w\frac{\partial w}{\partial x}\right) - \left(v\frac{\partial u}{\partial y} + w\frac{\partial u}{\partial z}\right)\\ &= v\zeta - w\eta\end{aligned}$$

and similarly

$$\left\{\begin{aligned}\frac{\partial E}{\partial x} + \frac{\partial u}{\partial t} &= v\zeta - w\eta\\ \frac{\partial E}{\partial y} + \frac{\partial v}{\partial t} &= w\xi - u\zeta\\ \frac{\partial E}{\partial z} + \frac{\partial w}{\partial t} &= u\eta - v\xi\end{aligned}\right\}.$$

(These are equivalent to the vector formula

$$\operatorname{grad} E + \frac{\partial \mathbf{v}}{\partial t} = \mathbf{v} \times \boldsymbol{\xi}.)$$

Now we consider the differential form

$$\omega = u\,dx + v\,dy + w\,dz - E\,dt.$$

We have

$$\begin{aligned}d\omega = {} & (\xi\,dy\,dz + \eta\,dz\,dx + \zeta\,dx\,dy)\\ & + dt\,(u_t\,dx + v_t\,dy + w_t\,dz)\\ & - (E_x\,dx + E_y\,dy + E_z\,dz)\,dt,\end{aligned}$$

$$\begin{aligned}d\omega = {} & (\xi\,dy\,dz + \eta\,dz\,dx + \zeta\,dx\,dy)\\ & + dt[(v\zeta - w\eta)\,dx + (w\xi - u\zeta)\,dy + (u\eta - v\xi)\,dz].\end{aligned}$$

Actually, $d\omega$ is an integral-invariant so that ω is a relative integral-invariant. One sees this indirectly by making a comparison to Hamiltonian systems.

For if one thinks for the moment of x, y, z, u, v, w, t as independent variables, the equations of motion are

$$\begin{Bmatrix} \dot{x} = u & \dot{u} = -\dfrac{\partial}{\partial x}(V + q) \\ \dot{y} = v & \dot{v} = -\dfrac{\partial}{\partial y}(V + q) \\ \dot{z} = w & \dot{w} = -\dfrac{\partial}{\partial z}(V + q) \end{Bmatrix}$$

or

$$\begin{Bmatrix} \dot{x} = \partial E/\partial u & \dot{u} = -\partial E/\partial x \\ \dot{y} = \partial E/\partial v & \dot{v} = \partial E/\partial y \\ \dot{z} = \partial E/\partial w & \dot{w} = -\partial E/\partial z \end{Bmatrix}$$

which makes our assertion clear. It follows that if $\mathbf{c}_2$, $\mathbf{c}'_2$ are two-chains in $(t, \mathbf{x})$ space which are in one-one correspondence so that corresponding points are on the same trajectory, then

$$\int_{\mathbf{c}_2} d\omega = \int_{\mathbf{c}'_2} d\omega.$$

In particular if $\mathbf{c}_2^{(0)}$ is a 2-chain in $\mathbf{E}^3$ at time t_0 and it moves to $\mathbf{c}_2^{(1)}$ at time t_1 according to the motion, then

$$\int_{\mathbf{c}_2^{(0)}} (\xi\, dy\, dz + \eta\, dz\, dx + \zeta\, dx\, dy) = \int_{\mathbf{c}_2^{(1)}} (\xi\, dy\, dz + \eta\, dz\, dx + \zeta\, dx\, dy).$$

This is the Helmholtz theorem on *conservation of vorticity*. In the first integral we must understand $\xi = \xi(t_0, \mathbf{x})$, etc., and in the second, $\xi = \xi(t_1, \mathbf{x})$, etc. An important consequence is this further result of Helmholtz. *Suppose at a fixed time t_0, the vorticity vanishes identically. Then it always vanishes.* For by the formula above,

$$\int_{\mathbf{c}_2^{(1)}} (\xi\, dy\, dz + \cdots) = 0$$

for each 2-chain $\mathbf{c}_2^{(1)}$ at each time t_1, which evidently means that the integrand must vanish.

10.7. Problems

1. On p. 54 and again on pp. 179–180 we defined a vector field $\mathbf{v}$ on a manifold $\mathbf{M}$ as a mapping which takes each function f on $\mathbf{M}$ to another function $\mathbf{v}(f)$ and satisfies the rules

$$\mathbf{v}(f + g) = \mathbf{v}(f) + \mathbf{v}(g)$$
$$\mathbf{v}(fg) = f \cdot \mathbf{v}(g) + \mathbf{v}(f) \cdot g.$$

We know that locally

$$\mathbf{v} = \sum a^i(x) \frac{\partial}{\partial x^i}.$$

We have alluded several times to the duality between vector fields and one-forms. (Cf. pp. 127, 143 for example.) Now we shall make this more explicit by associating with each vector field $\mathbf{v}$ and each one-form σ on $\mathbf{M}$ a scalar $(\sigma, \mathbf{v})$. Locally, if $\mathbf{v}$ has the representation above and

$$\sigma = \sum b_i(x)\, dx^i,$$

then we set

$$(\sigma, \mathbf{v}) = \sum a^i(x) b_i(x).$$

Show that

$$(\sigma_1 + \sigma_2, \mathbf{v}) = (\sigma_1, \mathbf{v}) + (\sigma_2, \mathbf{v}),$$

$$(\sigma, \mathbf{v}_1 + \mathbf{v}_2) = (\sigma, \mathbf{v}_1) + (\sigma, \mathbf{v}_2),$$

$$(g\sigma, \mathbf{v}) = g \cdot (\sigma, \mathbf{v}) = (\sigma, g\mathbf{v})$$

for each scalar g. Show also that this operation has the intrinsic characterization

$$(dg, \mathbf{v}) = \mathbf{v}(g).$$

2. Show similarly that two-forms may be paired with two-vectors by

$$(\sigma_1 \wedge \sigma_2, \mathbf{v}_1 \wedge \mathbf{v}_2) = \begin{vmatrix} (\sigma_1, \mathbf{v}_1) & (\sigma_1, \mathbf{v}_2) \\ (\sigma_2, \mathbf{v}_1) & (\sigma_2, \mathbf{v}_2) \end{vmatrix}.$$

3. Now prove the formula

$$(d\sigma, \mathbf{v} \wedge \mathbf{w}) + (\sigma, [\mathbf{v}, \mathbf{w}]) = \mathbf{v}\{(\sigma, \mathbf{w})\} - \mathbf{w}\{(\sigma, \mathbf{v})\},$$

which relates all these things to the Lie bracket.

4. Combine this with the Frobenius integration theorem (Section 7.3) to establish this result: Let $\mathbf{v}_1, \cdots, \mathbf{v}_r$ be vector fields on a neighborhood of 0 in $\mathbf{E}^n$ which are linearly independent at each point. Suppose that for each i and j, $[\mathbf{v}_i, \mathbf{v}_j]$ is a linear combination of $\mathbf{v}_1, \cdots, \mathbf{v}_r$. Then there is a coordinate system $x^1, \cdots, x^n$ of some neighborhood of 0 such that

$$\mathbf{v}_i = \sum_{j=1}^{r} a_i{}^j(x) \frac{\partial}{\partial x^j} \qquad (i = 1, \cdots, r).$$

5. Let $\mathbf{v}_1, \cdots, \mathbf{v}_r$ be vector fields on a neighborhood of 0 in $\mathbf{E}^n$ which are linearly independent at each point. Suppose that each bracket $[\mathbf{v}_i, \mathbf{v}_j]$ vanishes. Prove that there exists a local coordinate system $x^1, \cdots, x^n$ such that

$$\mathbf{v}_i = \frac{\partial}{\partial x^i} \qquad (i = 1, \cdots, r).$$

6. Let $u_1, \cdots, u_n$ be functions of $x^1, \cdots, x^n$ and set

$$J = \|\partial u_i/\partial x^j\|,$$

the Jacobian matrix. Write $J = A + B$ where A is symmetric and B is skew so that $A = \|a_{ij}\|$, $B = \|b_{ij}\|$

$$a_{ij} = \frac{1}{2}\left(\frac{\partial u_i}{\partial x^j} + \frac{\partial u_j}{\partial x^i}\right), \qquad b_{ij} = \frac{1}{2}\left(\frac{\partial u_i}{\partial x^j} - \frac{\partial u_j}{\partial x^i}\right).$$

(i) Suppose $B = 0$. Prove there is a function f satisfying $df = \sum u_i\, dx^i$.

(ii) Suppose $A = 0$. Prove that J is constant so that the u_i are linear functions of $x^1, \cdots, x^n$. (This comes up in the theory of strain.)

7. Let a_{ij} be n^2 functions of $x^1, \cdots, x^n$. Show that the integrability conditions

$$\frac{\partial^2 a_{ij}}{\partial x^k\, \partial x^l} - \frac{\partial^2 a_{ik}}{\partial x^j\, \partial x^l} = \frac{\partial^2 a_{lj}}{\partial x^k\, \partial x^i} - \frac{\partial^2 a_{lk}}{\partial x^j\, \partial x^i}$$

are necessary and sufficient for the existence of functions $u_1, \cdots, u_n$ satisfying

$$a_{ij} = \frac{1}{2}\left(\frac{\partial u_i}{\partial x^j} + \frac{\partial u_j}{\partial x^i}\right).$$

Investigate $n = 3$ and $n = 2$. (Due to Saint-Venant. See Love [18, p. 49].)

8. Let $\mathbf{v}$ be a vector field in $\mathbf{E}^3$. Show that there exists a function ϕ and a vector field $\mathbf{w}$ with $\operatorname{div}(\mathbf{w}) = 0$ so that

$$\mathbf{v} = \operatorname{grad}(\phi) + \operatorname{curl}(\mathbf{w}).$$

(*Hint*: use the mechanism of harmonic differentials. Assume that if g is any function in $\mathbf{E}^3$, the equation $\operatorname{Lap}(f) = g$ has a solution. See Abraham-Becker [1, pp. 37–38] and Love [18, p. 47].)

9. Consider the transformation

$$\bar{q}^i = \bar{q}^i(q^1, \cdots, q^n)$$

$$\bar{p}_i = \sum p_j \frac{\partial q^j}{\partial \bar{q}^i}.$$

The problem is to give a new proof of the relation

$$\frac{\partial \bar{p}_i}{\partial q^k} + \frac{\partial p_k}{\partial \bar{q}^i} = 0$$

derived in Section 10.1. Set

$$\lambda_j = dp_j - \sum \frac{\partial \bar{q}^i}{\partial q^j} d\bar{p}_i$$

and prove

(i) $\sum \lambda_j \wedge dq^j = 0$

(ii) $\lambda_j = \sum g_{jk} dq^k$

from which you conclude that $g_{jk} = g_{kj}$. Show that this is equivalent to the desired relations.

Bibliography

There is no attempt at completeness nor historical accuracy. In general we have tried to avoid references to less readily accessible material.

The most elementary introduction to our subject is found in [5]. More advanced introductions will be found in [7], [12], [13], and [19]; the algebraic aspects are treated fully in [4]. Items [2], [20], and to some degree [30] supply all one needs in topology. The principal sources of our material on physics are [1], [8], [14], [17], [18], [21], and [22]. Further applications of exterior differential form theory to physical science will be found in [25], [27], [28], [29], and [31]. An important source for material on differential equations is [6].

Advanced material on Lie groups is contained in [10], [11], and [16], and on differential geometry in [3], [7], [8], [9], [10], [12], [16], [23], [24], and [26].

I. Books

1. Abraham, M., and Becker, R., *Classical theory of electricity and magnetism*, 2nd ed., Glasgow (1950).
2. Alexandroff, P., and Hopf, H., *Topologie*, Berlin (1935), esp. pp. 465–7 and 497–8.
3. Blaschke, W., *Einführung in die Differentialgeometrie*, Berlin (1950).
4. Bourbaki, N., *Algèbre*, Chapitre III, *Algèbre multinéaire*, Act. Sci. et Ind. 1044, Paris (1948).
5. Buck, R. C., *Advanced calculus*, New York (1956), esp. pp. 309–21.
6. Carathéodory, C., *Variationsrechnung und partielle Differentialgleichungen erster Ordnung*, Leipzig (1935), reprinted Ann Arbor (1945).
7. Cartan, É., *Leçons sur la géométrie des espaces de Riemann*, Paris (1946).
8. Cartan, É., *Leçons sur les invariants integraux*, Paris (1921), reprinted (1958).
9. Cartan, É., *Systèmes différentiels extérieurs et leurs applications géométriques*, Act. Sci. et Ind. 994, Paris (1945).
10. Cartan, É., *Théorie des groupes finis et continus et la géométrie différentielle traitées par la méthode du repère mobile*, Paris (1951).
11. Chevalley, C., *Theory of Lie groups* I, Princeton (1946).
12. De Rham, G., *Variétés différentiables, formes, courants, formes harmoniques*, Paris (1955).

13. Goldberg, S. I., *Curvature and homology*, New York (1962).
14. Golomb, M., *Theoretical mechanics*, Purdue lecture notes, Lafayette (1960), esp. pp. 105–115.
15. Goursat, É., *Course in Mathematical Analysis*, Vol. I, Boston (1904).
16. Hodge, W. V. D., *Theory and applications of harmonic integrals*, Cambridge (1952).
17. Lamb, Horace, *Hydrodynamics*, 6th ed., Cambridge (1932), reprinted New York (1945).
18. Love, A. E. H., *Treatise on the mathematical theory of elasticity*, 4th ed., Cambridge (1927), reprinted New York (1944).
19. Nickerson, H. K., Spencer, D. C., and Steenrod, N., *Advanced calculus*, Preliminary edition, Princeton (1959).
20. Seifert, H., und Threlfall, W., *Lehrbuch der Topologie*, Leipzig (1934), reprinted New York (1947).
21. Sommerfeld, A. J. W., *Electrodynamics, Lectures on theoretical physics*, Vol. 3, New York (1952).
22. Whittaker, E. T., *Treatise on the analytical dynamics of particles and rigid bodies*, 4th ed., Cambridge (1937), reprinted New York (1944).
23. Willmore, T. J., *Introduction to differential geometry*, Oxford (1959), esp. pp. 189–192 and 202–205.

II. Papers

24. Chern, S. S., *Some new viewpoints in differential geometry in the large*, Bull. Amer. Math. Soc. **52** (1946), pp. 1–30.
25. Debever, R., *Espaces de l'electromagnétisme*, Colloque de géométrie différentielle, Louvain (1951), pp. 217–233.
26. Flanders, H., *Development of an extended exterior differential calculus*, Trans. Amer. Math. Soc. **75** (1953), pp. 311–326.
27. Gallissot, F., *Application des formes extérieures du 2e ordre à la dynamique Newtonienne et relativiste*, Annals de l'institut Fourier **3** (1951), pp. 277–285.
28. Gallissot, F., *Formes extérieures en mécanique*, ibid. **4** (1952), pp. 145–297.
29. Gallissot, F., *Formes extérieures et la mécanique des milleux continus*, ibid. **8** (1958), pp. 291–335.
30. Hadamard, J., *Sur quelques applications de l'indice de Kronecker*, appended to *J. Tannery, Introduction à la théorie des fonctions d'une variable* II, 2nd ed., Paris (1910).
31. Misner, C. W., and Wheeler, J. A., "Gravitation, electromagnetism, unquantized charge, and mass as properties of curved empty space," Ann. of Physics **2** (1957), pp. 525–603. Reprinted in *Geometrodynamics*, Topics in Modern Physics I, Academic Press, 1962, pp. 225–307.

III. Addendum to the Dover Edition

32. Flanders, H., "Differential Forms," Chapter 4 in Chern, S. S., editor, *Studies in Global Geometry and Analysis*, Mathematical Association of America (1967), pp. 57–95. The proof of the converse on Poincaré's lemma, page 66, will be interesting.

33. Flanders, H., "Differential Forms," Chapter 2 in Chern, S. S., editor, *Global Differential Geometry*, MMA Studies in Mathematics, No. 27, Mathematical Association of America, 1989, pp. 27–72. Please note that this text is an updated, retitled edition of the text cited immediately above. This contains three items of interest. First, there is still another inductive and very clean proof of the converse of Poincaré's lemma (Section 3). Second, it contains an exposition of Y. Kannai's proof of the Brouwer fixed-point theorem (Section 7).

Glossary of Notation

This is a list of special terminology, definitions, and symbols, especially from the early chapters.

A. Spaces

$\mathbf{E}^n$ Euclidean n-space.

$\mathbf{R}$ The set of real numbers, also considered as $\mathbf{E}^1$, the Euclidean line.

$\mathbf{U}, \mathbf{V}, \cdots$ Open sets (in $\mathbf{E}^n$, or on a manifold).

$\mathbf{L}, \mathbf{M}, \cdots$ Vector spaces.

$\bigwedge^p \mathbf{L}$ The space of p-vectors on $\mathbf{L}$.

$\mathbf{M}, \mathbf{N}, \cdots$ Manifolds.

$\mathbf{F}^p(\mathbf{U})$ The collection of all p-forms on $\mathbf{U}$.

Cartesian product. If $\mathbf{S}$ and $\mathbf{T}$ are arbitrary sets (collections of objects), their *cartesian product* is the set

$$\mathbf{S} \times \mathbf{T}$$

consisting of all ordered pairs (s, t) where s belongs to $\mathbf{S}$ and t to $\mathbf{T}$.

$\times^2 \mathbf{S}$ This is the cartesian product $\mathbf{S} \times \mathbf{S}$ of $\mathbf{S}$ with itself. Similarly

$$\times^3 \mathbf{S} = \mathbf{S} \times \mathbf{S} \times \mathbf{S}, \qquad \text{etc.}$$

$\mathbf{S} \cap \mathbf{T}$ This is the *intersection* of the sets $\mathbf{S}$ and $\mathbf{T}$. For example, if $\mathbf{S} = \{1, 2, 5, 7\}$ and $\mathbf{T} = \{2, 3, 7, 9\}$, then $\mathbf{S} \cap \mathbf{T} = \{2, 7\}$.

$\mathbf{I}$ The unit interval $0 \leqq t \leqq 1$.

B. Functions

Mapping. A *mapping* is a smooth function ϕ from one space $\mathbf{M}$ to another $\mathbf{N}$. We write

$$\phi\colon\ \mathbf{M} \longrightarrow \mathbf{N}.$$

Composite mapping. If $\phi\colon\ \mathbf{M} \longrightarrow \mathbf{N}$ and $\psi\colon\ \mathbf{N} \longrightarrow \mathbf{P}$, then we may form the *composite mapping* $\psi \circ \phi\colon\ \mathbf{M} \longrightarrow \mathbf{P}$. It is defined by

$$(\psi \circ \phi)(x) = \psi[\phi(x)] \quad \text{for } x \text{ in } \mathbf{M}.$$

See pp. 3, 22, ff.

Linear functional. A linear transformation on a linear space **L** to the one-dimensional space **R** of real numbers.

Jacobian. If $u^i = u^i(x^1, \cdots, x^n)(i = 1, \cdots, n)$, the *Jacobian* of this mapping is the determinant

$$|\partial u^i/\partial x^j|.$$

ϕ^* The mapping on differential forms induced by the mapping ϕ between spaces, p. 23, ff.

ϕ_* The mapping on chains induced by the mapping ϕ between spaces, p. 71, ff.

C. Special symbols

${}^tA$ The transpose of the matrix A, obtained from A by interchanging rows and columns.

$|A|$ The determinant of the linear transformation (matrix) A, p. 21, ff.

dim **L** The dimension of the linear space **L**.

σ^H Here $H = \{h_1, \cdots, h_p\}$, a set of indices in increasing order,

$$1 \leqq h_1 < h_2 < \cdots < h_p \leqq n, \qquad \text{and} \qquad \sigma^H = \sigma^{h_1} \wedge \sigma^{h_2} \wedge \cdots \wedge \sigma^{h_p}.$$

H' This is the complementary set of indices. For example, if $n = 8$ and $H = \{2, 3, 5, 6\}$, then $H' = \{1, 4, 7, 8\}$.

$\operatorname{sgn} \pi$ If π is a permutation on $\{1, 2, \cdots, n\}$, then $\operatorname{sgn} \pi = 1$ if π is effected by an even number of interchanges (of two numbers) and $\operatorname{sgn} \pi = -1$ if π is effected by an odd number of interchanges.

$*$ The star operator, p. 15.

(α, β) The inner product, p. 12, ff.

∂ The boundary operator, p. 58.

$$dx^1 \wedge \cdots \wedge \widehat{dx^i} \wedge \cdots \wedge dx^n \text{ means } dx^1 \wedge \cdots \wedge dx^{i-1} \wedge dx^{i+1} \wedge \cdots \wedge dx^n.$$

The circumflex indicates omission.

Index

A CATALOG OF SELECTED

DOVER BOOKS

IN SCIENCE AND MATHEMATICS

SPECIAL FUNCTIONS, N.N. Lebedev. Translated by Richard Silverman. Famous Russian work treating more important special functions, with applications to specific problems of physics and engineering. 38 figures. 308pp. 5⅜ × 8½. 60624-4 Pa. $6.95

OBSERVATIONAL ASTRONOMY FOR AMATEURS, J.B. Sidgwick. Mine of useful data for observation of sun, moon, planets, asteroids, aurorae, meteors, comets, variables, binaries, etc. 39 illustrations 384pp. 5⅜ × 8¼. (Available in U.S. only) 24033-9 Pa. $5.95

INTEGRAL EQUATIONS, F.G. Tricomi. Authoritative, well-written treatment of extremely useful mathematical tool with wide applications. Volterra Equations, Fredholm Equations, much more. Advanced undergraduate to graduate level. Exercises. Bibliography. 238pp. 5⅜ × 8½. 64828-1 Pa. $6.95

CELESTIAL OBJECTS FOR COMMON TELESCOPES, T.W. Webb. Inestimable aid for locating and identifying nearly 4,000 celestial objects. 77 illustrations. 645pp. 5⅜ × 8½. 20917-2, 20918-0 Pa., Two-vol. set $12.00

MODERN NONLINEAR EQUATIONS, Thomas L. Saaty. Emphasizes practical solution of problems; covers seven types of equations. ". . . a welcome contribution to the existing literature. . . ."—*Math Reviews*. 490pp. 5⅜ × 8½. 64232-1 Pa. $9.95

FUNDAMENTALS OF ASTRODYNAMICS, Roger Bate et al. Modern approach developed by U.S. Air Force Academy. Designed as a first course. Problems, exercises. Numerous illustrations. 455pp. 5⅜ × 8½. 60061-0 Pa. $8.95

INTRODUCTION TO LINEAR ALGEBRA AND DIFFERENTIAL EQUATIONS, John W. Dettman. Excellent text covers complex numbers, determinants, orthonormal bases, Laplace transforms, much more. Exercises with solutions. Undergraduate level. 416pp. 5⅜ × 8½. 65191-6 Pa. $8.95

INCOMPRESSIBLE AERODYNAMICS, edited by Bryan Thwaites. Covers theoretical and experimental treatment of the uniform flow of air and viscous fluids past two-dimensional aerofoils and three-dimensional wings; many other topics. 654pp. 5⅜ × 8½. 65465-6 Pa. $14.95

INTRODUCTION TO DIFFERENCE EQUATIONS, Samuel Goldberg. Exceptionally clear exposition of important discipline with applications to sociology, psychology, economics. Many illustrative examples; over 250 problems. 260pp. 5⅜ × 8½. 65084-7 Pa. $6.95

LAMINAR BOUNDARY LAYERS, edited by L. Rosenhead. Engineering classic covers steady boundary layers in two- and three-dimensional flow, unsteady boundary layers, stability, observational techniques, much more. 708pp. 5⅜ × 8½. 65646-2 Pa. $15.95

LECTURES ON CLASSICAL DIFFERENTIAL GEOMETRY, Second Edition, Dirk J. Struik. Excellent brief introduction covers curves, theory of surfaces, fundamental equations, geometry on a surface, conformal mapping, other topics. Problems. 240pp. 5⅜ × 8½. 65609-8 Pa. $6.95

ROTARY-WING AERODYNAMICS, W.Z. Stepniewski. Clear, concise text covers aerodynamic phenomena of the rotor and offers guidelines for helicopter performance evaluation. Originally prepared for NASA. 537 figures. 640pp. 6⅛ × 9¼.
64647-5 Pa. $14.95

DIFFERENTIAL GEOMETRY, Heinrich W. Guggenheimer. Local differential geometry as an application of advanced calculus and linear algebra. Curvature, transformation groups, surfaces, more. Exercises. 62 figures. 378pp. 5⅜ × 8½.
63433-7 Pa. $7.95

INTRODUCTION TO SPACE DYNAMICS, William Tyrrell Thomson. Comprehensive, classic introduction to space-flight engineering for advanced undergraduate and graduate students. Includes vector algebra, kinematics, transformation of coordinates. Bibliography. Index. 352pp. 5⅜ × 8½. 65113-4 Pa. $8.00

A SURVEY OF MINIMAL SURFACES, Robert Osserman. Up-to-date, in-depth discussion of the field for advanced students. Corrected and enlarged edition covers new developments. Includes numerous problems. 192pp. 5⅜ × 8½.
64998-9 Pa. $8.00

ANALYTICAL MECHANICS OF GEARS, Earle Buckingham. Indispensable reference for modern gear manufacture covers conjugate gear-tooth action, gear-tooth profiles of various gears, many other topics. 263 figures. 102 tables. 546pp. 5⅜ × 8½. 65712-4 Pa. $11.95

SET THEORY AND LOGIC, Robert R. Stoll. Lucid introduction to unified theory of mathematical concepts. Set theory and logic seen as tools for conceptual understanding of real number system. 496pp. 5⅜ × 8¼. 63829-4 Pa. $8.95

A HISTORY OF MECHANICS, René Dugas. Monumental study of mechanical principles from antiquity to quantum mechanics. Contributions of ancient Greeks, Galileo, Leonardo, Kepler, Lagrange, many others. 671pp. 5⅜ × 8½.
65632-2 Pa. $14.95

FAMOUS PROBLEMS OF GEOMETRY AND HOW TO SOLVE THEM, Benjamin Bold. Squaring the circle, trisecting the angle, duplicating the cube: learn their history, why they are impossible to solve, then solve them yourself. 128pp. 5⅜ × 8½. 24297-8 Pa. $3.95

MECHANICAL VIBRATIONS, J.P. Den Hartog. Classic textbook offers lucid explanations and illustrative models, applying theories of vibrations to a variety of practical industrial engineering problems. Numerous figures. 233 problems, solutions. Appendix. Index. Preface. 436pp. 5⅜ × 8½. 64785-4 Pa. $8.95

CURVATURE AND HOMOLOGY, Samuel I. Goldberg. Thorough treatment of specialized branch of differential geometry. Covers Riemannian manifolds, topology of differentiable manifolds, compact Lie groups, other topics. Exercises. 315pp. 5⅜ × 8½. 64314-X Pa. $6.95

HISTORY OF STRENGTH OF MATERIALS, Stephen P. Timoshenko. Excellent historical survey of the strength of materials with many references to the theories of elasticity and structure. 245 figures. 452pp. 5⅜ × 8½. 61187-6 Pa. $9.95